Springer Series in Design and Innovation

Volume 68

Springer Series in Design and Innovation (SSDI) publishes books on innovation and the latest developments in the fields of Product Design, Interior Design and Communication Design, with particular emphasis on technological and formal innovation, and on the application of digital technologies and new materials. The series explores all aspects of design, e.g. Human-Centered Design/User Experience, Service Design, and Design Thinking, which provide transversal and innovative approaches oriented on the involvement of people throughout the design development process. In addition, it covers emerging areas of research that may represent essential opportunities for economic and social development.

In fields ranging from the humanities to engineering and architecture, design is increasingly being recognized as a key means of bringing ideas to the market by transforming them into user-friendly and appealing products or services. Moreover, it provides a variety of methodologies, tools and techniques that can be used at different stages of the innovation process to enhance the value of new products and services.

The series' scope includes monographs, professional books, advanced textbooks, selected contributions from specialized conferences and workshops, and outstanding Ph.D. theses.

The volumes of the series are single-blind peer-reviewed.

Keywords: Product and System Innovation; Product design; Interior design; Communication Design; Human-Centered Design/User Experience; Service Design; Design Thinking; Digital Innovation; Innovation of Materials.

How to submit proposals

Proposals must include: title, keywords, presentation (max 10,000 characters), table of contents, chapter abstracts, editors'/authors' CV.

In case of proceedings, chairmen/editors are requested to submit the link to conference website (incl. relevant information such as committee members, topics, key dates, keynote speakers, information about the reviewing process, etc.), and approx. number of papers.

Proposals must be sent to publishing editor Mr. Pierpaolo Riva (pierpaolo.riva@springer.com).

Vivek Kant
Editor

Human Centred Design for Systems

Springer

Editor
Vivek Kant
Department of Design
Indian Institute of Technology Kanpur
Kanpur, Uttar Pradesh, India

ISSN 2661-8184 ISSN 2661-8192 (electronic)
Springer Series in Design and Innovation
ISBN 978-3-032-30375-2 ISBN 978-3-032-30376-9 (eBook)
https://doi.org/10.1007/978-3-032-30376-9

This Springer imprint is published by the registered company Springer Nature Switzerland AG
The registered company address is: Gewerbestrasse 11, 6330 Cham, Switzerland

Contents

Chapter 1
Introduction: The Way Forward with Human Centered Design for Systems

Vivek Kant

Abstract Systems are rapidly increasing in complexity. One of the challenges in our times is to handle the complexity in a meaningful manner in order to make the tractable from a human-centric viewpoint. In order to continue the way forward for a human-centric approach to systems, there is a need to address three inter-related dimensions of systems approach: Culture of Care, Systems Practice, Reflective Engagement. These three dimensions together will enable a human-centric approach to systems design will shape the systems of tomorrow to become more human-centric.

Keywords Complex systems · Culture of care · Systems practice · Reflective engagement · Human-Centered design

1.1 What Are Systems?

If we look around, we find many things around us in day-to-day life, right from the invisible cells in the human body to vast, unknown constellations in space. Using the concept of a system, helps to comprehend the multitude of Interactions and behaviors of these entities. Let's consider spaces as an example, to understand the whole concept of systems in a much better and easier way. The universe can be considered as a system. Further within the universe as a system we have several subsystems within it, such as constellations. Similarly, these constellations and galaxies can be seen In Isolation as a system too. Then under galaxies we can find one specific galaxy called the Milky Way considered as a subsystem, we can comprehend Solar System as one and then under Solar system we can focus on Earth and further delve deeper into the various system that are present within Earth.

The example of space is provided to recognize that systems are everywhere, and these Systems are complex and constantly evolving! In a sense, these systems can be considered as some kind of interdependent assembly of elements that function together for a common objective. In other words, It can be understood as the whole and

V. Kant (✉)
Indian Institute of Technology Kanpur, Kanpur, India
e-mail: vkant@iitk.ac.in

V. Kant (ed.), *Human Centred Design for Systems*, Springer Series in Design and Innovation 68, https://doi.org/10.1007/978-3-032-30376-9_1

its parts, where the parts work together to form a whole and are considered of various Systems exist at several different levels. Physical, social, digital and ecological, many other among there is a constant exchange within them be it of various aspects dimension such as information, energy or matter.

Every system has a few key dimensions such as structure (how it's built), function (what it does), and feedback (how it adapts). In design, in particular, the recognition of systems shifts the focus from isolated artifacts to relationships, flows, and their interactions within. As a chair does not exist as a product alone but can fit into various broader categories, while simultaneously existing as standalone entities, that fit in with the overall interior and aesthetic of the room, as part of ergonomics, as part of material supply, labor and craftsmanship, culture, and even within sustainability. In the same way, a service is not just an isolated process but akin to a network of people, tools, and values. Systems thinking thus helps not only the designers but anyone who wants to understand the system to move from linear cause-and-effect thinking to circular, dynamic, and adaptive understanding of how things work together within it. Understanding the system allows one to know what type of system it is, the importance of key elements in it, whether its a social system, technical system or a socio-technical system and how small changes within these systems can lead to large-scale side effects. In general, these systems are complicated and messy in the sense, oversimplification of its complexity and solutions or changes without understanding its causes can result in ripple effects that are unpredictable and can go across the whole system in nonlinear ways.

In a time when problems have become increasingly "wicked"—Challenges such as climate change, pollution, population, sustainability, inequality, digital addiction, and overconsumption have become ever-evolving with increased complexity and no immediate solutions at hand [1, 2]. This is where systems thinking offers a lens to design responsibly, providing the need to explore its connections over components, processes over products, and emergence over control. Systems are not just the outer lying things we design for, but they are the things we design within and sometimes, design with. Understanding these systems thus forms the backbone of designing for complexity, resilience, and meaningful change in the modern world where all the connections are given priority and established well within.

1.2 Why Is this Book Important Now?

We exist in a world that is moving at a fast pace where innovations happen almost every day for improving lives. But if we look at it, these innovations bring a lot of changes in everyday lives, impacting many. We can observe that all these systems are all interdependent and complex at the same time, be they technological, social, or ecological, where each system is interconnected in ways like never before and fast evolving. Every design decision made today, be how small it may, has its effects that reach far beyond its context. Today's designers experience problems that simply cannot be addressed or resolved by analyzing how something looks, feels, or works.

The need to understand its reciprocations and repulsions are greater than before, as one needs a much broader, wider view of the people, processes, causes, or influences shaping the behaviors, and their outcomes are crucial at a systems level.

Now is the time, as the design of things is moving into the design of relationships at a broader level [3]. This Is the crucial level where the importance of this book is stressed upon to understand those interconnected relationships and address them holistically. The role of today's designer is evolving rapidly along with the rise of technology and automation as the designers transform from being makers to more of strategic thinkers, facilitators, and changemakers [4–6]. Systems design empowers them to understand the invisible architectures of interaction that govern both the human and environmental well-being in the long run.

In a world where people have started to question the underlying assumptions of environmental sustainability (Planet), Social Sustainability (People), Economic Sustainability (Profit), this book will act as a step toward rethinking the general approach to design, extending beyond problem-solving, towards problem-framing, anticipation and adaptation. Drawing on existing literature, it decodes the underlying elements that define modern challenges, enabling design that's not just mere functional but systemic, ethical and future-proof in all ways!

1.3 Why a Book on HCD for Systems?

If we look at systems alone, we see a complex web-like structure that brings into account all the interconnections and hidden underlying factors, but when combined with the human aspect, it can bring a lot more value for the designers to look at and intervene in the most necessary areas. Human-Centered Design has long inspired designers to design with empathy, to understand the users, their pain points and the aspirations and an output that is something that benefits the user. As systems grow in both breadth and width, increasing the complexity, the user that is being addressed is no longer a user as an individual per say. Its more of a web of stakeholders each influencing and being influenced by design in their own way. This shift calls for an evolution from Human-centered design into Humanity-centered and ecological-centred systems level design taking into consideration of the whole [1, 2].

The holistic emphasis of this book is to bring the aspect of both HCD and Systems Thinking together as both share a common philosophy at its core, emphasizing on "understanding the context before any action." As HCD provides us the empathy to connect to people, Systems Thinking gives the much needed perspective, tools and direction to understand the environments shaping them. Together they form a framework that is humane in value and systemic in process.

This framework of HCD for Systems ensures that the decisions taken at any level are situated well within the everyday reality and translates concepts and models to a human level, enabling the designer to visualize how the ripple effect for various design choices can translate even to the bottom.

This book using examples, is more about putting forward a way to rethink, access and question how the role and responsibility of a designer is more about shaping of systems that sustains rather than just applying the necessary tools and methods. Its all about keeping the human factor at the center of every system that we imagine, build and develop!

1.4 How Will It Help Designers?

This book acts for designers as both a guide and as a mirror of reflection. It guides designers through the complexity while mirroring back how their work can impact the systems at large. It helps one to understand the multiple levels of data and their connections as it acts as a tool for identifying gaps and potential opportunities in improving the overall system. It can be envisioned as a mental toolkit, demonstrating step by step methodologies to be followed in order to unravel and understand the system chosen. The book, explains in simple terms with examples in the form of diverse projects and through different perspectives on how to map out the systems, stakeholders involved, their relationships and their interconnectedness within them. While prominent texts such as *Designing Complexity: The Methodology and Practice of Systems Oriented Design* by *Birger Sevaldson* [5] and *Design Journeys through Complex Systems: Practice Tools for Systemic Design* by *Peter Jones, Kristel Van Ael* [4] exist, this current book extends their spirit through practical examples. It also emphasizes on ways of finding the potential, intervention areas, and even to access the proposed solution keeping in mind the long-term effects and reverse effects it can cause. It helps the designers to gain the ability to identify leverage points and design interventions that create meaningful, scalable and a sustainable change.

The book focuses on creating a shift of mindset from problem specific oriented to system-oriented thinking, helping to understand dynamic relationships between various parts of the system, adapt to framing problems at various levels within the systems and facilitate for collaboration beyond design and create a systemic innovation.

Designers are being trained to see the "whole" rather than the "parts" in an isolated form. Designers can now emphasize on creating outcomes that are adaptive, inclusive, and resilient. The systems lens will empower them to design with awareness of not just seeing what their design does but what impact it brings in the larger ecosystem. The methods described in this book, will help one to think more broadly, start asking "who is affected?" and "what shapes this issue?" and not just "how do I fix this right now?" This approach means the book not only helps to solve, but it also provides skills to envision and use in any design-oriented field where the challenges are complex and interconnected!

1.5 Structure of the Book

The book is divided into ten chapters where the first one lays out the foundation of Human Centered Designing for systems. The subsequent chapters use these concepts for developing various examples of Human Centered Design In systems. This ensures that the book Is focused primarily towards practicing designers who aim to imbibe principles of Human-centered Design In their projects. All of the projects developed in the book are from projects by designers and started as projects in a course on HCD for Systems. As a result, they emphasize the pedagogical dimension of HCD in systems.

Chapters 3–9 represent a number of projects where the core dimension has been presented as the *Relational Systems Design Conference (R&D 14)* at Ontario, Canada.[1] These chapters present an expanded and detailed version of these projects. Chapter 9 was extended further and published as a journal paper in *The Design Journa*[2]*l.*

Chapter 2 acts as the introduction to the Foundation of Human Centered Design for Systems, as it introduces key concepts in systems thinking and HCD. HCD can be applied to wicked problems that are characterized by complexity, uncertainty, and having multiple perspectives, all in myriad dimensions. It provides new thinking tools and concepts that help designers, students, and other thinkers make sense of complex systems, moving them from a linear cause-and-effect way of thinking towards a more holistic, circular, adaptive understanding of the overall system. The chapter outlines the evolution of systems thinking, establishes key concepts such system boundaries, metaphors, and relationships, and then describes in detail a five-step human-centred pathway to engaging with systems: framing, comprehending, envisioning, formulating, and enabling transition. It makes one understand the process and its importance how and why it is a need for every designer to start thinking at a systems level.

Chapter 3 is the first case study and discusses in detail about Disaster and Ethics, understanding the aftershocks of a disaster, specifically. This project closes the gap between logistical efficiency and moral responsibility in disaster management by studying the ethical dilemmas arising throughout earthquake relief. Ethical dilemmas were examined in three distinct phases: before, during, and after an earthquake. Then, the influence of decision-making and psychological stress on the choices made by the usual people involved was investigated. The most important intervention identified in this study was embedding ethical awareness in the preparedness phase and reframing ethics as proactive instead of reactive!

Chapter 4 is the second case study and introduction to Indian millets and their ecology, where the interventions are focused for a healthier future. This chapter reviews the decline of millets in the Indian diet despite their superior nutritional

[1] https://rsdsymposium.org/indian-institute-of-technology-kanpur-hfss-studio-maps/

[2] Sharma, A., Kant, V., & Malhotra, S. (2025). Designing a participatory-based permaculture toolkit for marginal Indian farmers to support sustainable farming. *The Design Journal*, 1–21. doi:10.1080/14606925.2025.2573718

and ecological benefits compared to other dominant grains such as wheat and rice. Through stakeholder mapping and Human-Centred system analysis, the study determined that low consumer awareness, prioritization of policy for subsidized grains, and the perception of millets as a "grain of the poor" contributed to their decline. The principal intervention identified is improving millet literacy by developing experiential stores to change public perception and behavior and shift consumers' focus toward millets for better health and environmental sustainability. The mapping techniques, thought process and decisions taken were explained briefly along with the clear explanation of the final idea.

Chapter 5 focuses on Mushrooms and health remediation. This study examines the fast-growing market for Indian oyster mushrooms and its value chain for potential innovative and sustainable applications. In particular, mycoremediation which is considered as the technique that uses fungi for cleaning environmental pollutants. The entire mushroom ecosystem was studied in depth right from cultivating mushrooms to using them in the industry to the biological and chemical properties enabling the mycelium network to break down complex pollutants and clean the air. Realizing that there are commercialization barriers with mycoremediation, one of the leverage points on the indoor air pollution, which led to the design of "Aeolix," a sustainable, bio-inspired air purifier where conventional filters are replaced by a mycelium-based alternative!

Chapter 6 focuses on supporting small-scale sellers and farmers in organic vegetable markets in India. This research addresses the systemic inefficiencies in the organic vegetable markets, where high post-harvest losses and exploitation by middlemen drastically reduce farmers revenues. The project used stakeholder mapping and ZIP analysis to dissect the social and economic structure of vegetable sales and identified the core problem as a middleman-reliant selling structure. The key intervention was determined to be restructuring of the vegetable market selling structure to enable direct farmer-to-consumer connect. The final proposed solution, TazaBazaar, is a WhatsApp™ based marketplace plugin that leverages the platform's high adoption rate among farmers to facilitate direct transactions, reduce food wastage, and provide farmers with better prices and better quality and reliant produce for the consumers as well.

Chapter 7 focuses on Endurance looking at enabling the endurance of athletes in specific. This is a timely project that addresses the key challenges and urgent need for scientifically supported endurance training and recovery strategies among athletes, who frequently face conditions such as overtraining and poor recovery protocols. A detailed systems analysis of athletic endurance identified recovery as one of the high-impact leverage points in preventing injury and enhancing performance. The response is an innovative design concept in an Athlete Suit with integrated soft sensor technology and adaptive yarn textiles that transform recovery and injury prevention. These suits will proactively track muscle stress in real time, trigger warnings before injury, and monitor temperature, acting as smart training companions that will enable athletes to safely push their limits and extend career longevity by leveraging both the technological advancements and enhanced data visualization.

Chapter 8 focuses on Human-Dog relationships in detail, focusing on building empathetic human-canine relationships. This project aims to improve the relationship between dogs and their owners, especially in high-stress conditions found in military services, which also contribute to separation anxiety and miscommunication with their handlers. The study related to dog cognition, behavior, and specific issues that service dogs and their handlers confront. The point of leverage was real-time insight into a dog's state regarding its emotions and physiological state. The final product designed is Chitta, a smart wearable vest integrated with a mobile app for monitoring vital signs, emotional signs, and behavioral patterns by means of haptic and audio feedback for discrete communication; thus, strengthening the bonding between the handler and the dog.

Chapter 9 focuses on supporting research on Plastic Degrading Micro-organisms. This project investigates plastic pollution and the promising yet under-researched field of plastic-degrading microbes, which represent a natural solution to persistent polymer waste. Considering the whole plastic lifecycle, this project applied gigamapping and ZIP analysis to find that the most strategic leverage point is the one regarding the general lack of research and data on the microbes that are available. The final design solution pitched was Hyphae Hatch, a portable, affordable, and modular mycelium incubator and research kit. It brings the the research process to the hands of many, thereby letting independent researchers, scientists, and curious individuals to create diverse data about plastic-eating fungi to eventually accelerate enzymatic degradation solutions for large-scale plastic bioremediation that can be introduced at large scale with careful analysis and systemic view.

Finally, Chap. 10 focused on permaculture and how participatory design and social innovation can bring a significant change in the way farmers look at their processes. This Chap. 10 addresses the gaps in knowledge and training that are holding back the marginal farmers in India from adopting sustainable agriculture such as permaculture. The project emphasizes the codesign of a Participatory Permaculture Toolkit through a long, multi-step process of co-creation workshops both with farmers and designers. This process made sure that the toolkit consisted of visual aids, interactive models, and farm mapping techniques that were locally relevant, user-friendly, and based on farmers knowledge. The outcome is a physical and digital catalyst for conversations which is not only teaching permaculture principles but also enabling peer-to-peer learning and community ownership and discussions, thus acting as a reference model that would help policymakers and NGOs enable transitions to sustainable farming in India. It further emphasizes the role of the designer being a facilitator for a design outcome rather than the creator of one!

1.6 The Road Ahead: *Inculcating Systems Practice, Fostering a Culture of Care, Enabling Reflective Engagement*

Design today stands at a crossroads. The boundaries between disciplines, systems, and scales are dissolving, demanding that designers think beyond the product or service into the interconnected realities that shape our actual world. This book invites designers to embrace that complexity and not just to control it, but to understand, collaborate with, and act within it more consciously.

The book highlights myriad examples where human centered design plays a vital role in framing a problem, identifying an intervention, formulating the solution, and transforming the system based on reconfigured practices. The challenges faced in each sector in the eight projects documented in the book were supported by the foundational concepts detailed in chapter one. The key idea is that future systems require a triad of systems practice, reflective engagement, and a culture of care. All these three are important to build a strong human-centric view of engagement with systems. The bigger challenge we now face is how to imbibe these concepts and practice and take them forward. Specifically, how systemic insights can be taken forward by readers in their own projects. In order to accomplish this, the key direction is to assimilate a mindset for human centric design in systems. Here, three inter-related dimensions of engagement are necessary for charting a way forward: Inculcating systems practice, Enabling Reflective Engagement, Empowering a Culture of Care (Fig. 1.1).

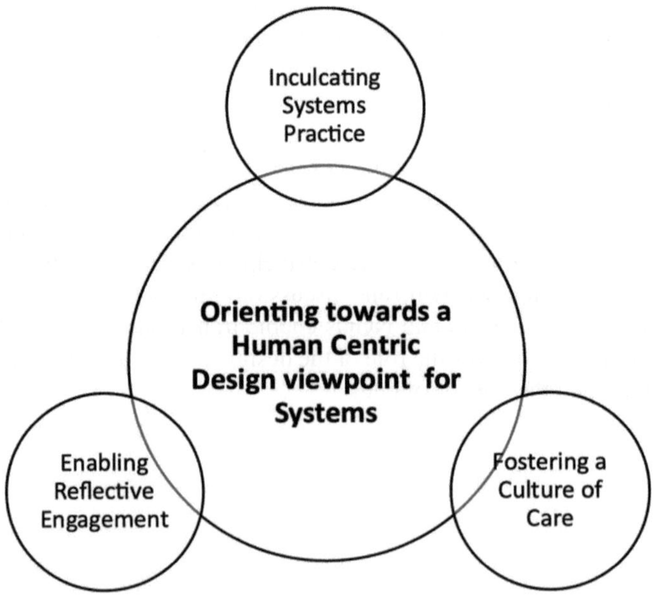

Fig. 1.1 Assimilate a mindset for human centric design in systems

Inculcating Systems Practice: This involves comprehending systems concepts and actively using them to guide everyday thinking and problem-solving [7, 8]. Systems concepts must be used in a way that shapes work practices to benefit the maximum number of stakeholders without being detrimental to the limited few within the system as a whole. This explicit thought process of integrating systems thinking into everyday design practice is an essential step towards the growth of human-centered design as a disciplinary practice for future practitioners. Adopting a human-centric mindset provides a distinct entry into the system, distinct from that of other stakeholders who engage with systems. Here, the designer adopts a viewpoint of effectiveness and engagement to ensure that the solutions are usable by stakeholders while considering the harmony of the system.

Fostering a Culture of Care: The second dimension of engagement that supports the first is fostering a culture of care [9, 10]. Here, the system focuses on developing a culture that accounts for stakeholders' meaning-making processes while recognizing the sustainability of the overall ecology. A culture of care takes into account the diverse experiences of stakeholders while valuing them as co-creators of innovations that improve the system as a whole. In this regard, there is a vital need for stakeholder engagement through participative processes. Participation of stakeholders emphasizes the collective understanding, shared awareness and inclusive co-creation of solutions that are acceptable for the designers and the end-users. Participation ensures that people involvement brings about a culture of trust as well as enables reflective engagements.

Enabling Reflective engagement: The final yet the most important dimension is that of enabling reflective engagement for designers as well as stakeholders [11]. The humans -centric processes enable the adoption of an empathetic viewpoint. This, in turn, opens up a new approach of comprehending the context from the other person's viewpoint. Such thought processes enable the co-participants in the systems design process to engage reflectively with its elements of the design process and its outcomes. This brings about a deeper and richer experience in creating human-centric systems.

The eight examples in the book demonstrate that, to support a human-centric view of design in systems, designers must be actively engaged in ethical systems practice. In order to achieve this, designers must imbibe the above three dimensions of *Inculcating Systems Practice, Fostering a Culture of Care, and Enabling Reflective engagement* in their design practice while engaging with systems. All these dimensions are mutually reinforcing; together, they enable human-centric design to flourish; taken together, they will shape the systems of tomorrow, while ensuring that designers play a pivotal role in systematic progression of systems.

Systems Thinking, along with Human Centered Design, moves us towards a more balanced approach, one that values empathy and structure along with the usage of visualization and analysis. This is not about finding answers within uncertainty but more about envisioning the unseen that are being hidden and left to be unraveled. In a view, systems design is all about the relationships, dependencies and feedback loops that reflects whether the design improves the harmony or causes harm overall. As the designers are now acting more as facilitators, this systemic mindset will evolve into shaping more resilient, equitable, and humane futures. This book isn't about a guide or acting as a manual but it stresses or emphasizes on a mindset shift. It asks one to stop, observe things and not design not for isolated problems or needs but for systems that sustains us all holistically!

References

1. Thackara J (2006) In the bubble: designing in a complex world. MIT press
2. Rittel HWJ, Webber MM (1973) Dilemmas in a general theory of planning. Policy Sci 4:155–169. https://doi.org/10.5040/9781474282932.0015
3. Golsby-Smith T (1996) Fourth order design: a practical perspective. Des Issues 12(1):5. https://doi.org/10.2307/1511742
4. Jones P, Van Ael K (2022) Design journeys through complex systems: practice tools for systemic design. BIS Publishers, Netherlands
5. Sevaldson B (2022) Designing complexity: the methodology and practice of systems oriented design. Common Ground Research Networks
6. Buchanan R (1992) Wicked problems in design thinking. Des Issues 8:5. https://doi.org/10.2307/1511637
7. Midgley G (2000) Systemic intervention. In: Systemic intervention: philosophy, methodology, and practice. Springer, Boston, MA, pp 113–133
8. Checkland P (1981) Systems thinking, systems practice. Wiley, Chichester
9. Jones P (2017) Social ecologies of flourishing: designing conditions that sustain culture. In: Design for a Sustainable Culture. Routledge, pp 38–54
10. Manzini E (2015) Design, when everybody designs: an introduction to design for social innovation. MIT Press
11. Escobar A (2018) Designs for the pluriverse: radical interdependence, autonomy, and the making of worlds. Duke University Press

Chapter 2
Foundations of Human Centered Design for Systems

Vivek Kant

Abstract Complex, interrelated systems are encountered everywhere. These natural, human-made, and socio-technical complex systems are highly interaction-driven and constantly evolving, creating unpredictability in their interactions. This chapter discusses how designers can address the unpredictable changes within a system using a holistic approach that moves beyond traditional linear problem-solving, which is required to wholly encompass the depth as well as the breadth of any system. This chapter introduces the concepts of wicked problems, drawing on the work of famous system theorists Horst Rittel, Melvin Weber, and Russell Ackoff to explain the characterization of systemic challenges, such as low agreement on problem definition, high unpredictability, and diverse perspectives of stakeholders. To deal with these challenges, a systems thinking approach equips designers with different perspectives and conceptual tools while enabling them to envision and represent a system's scope and constraints. These tools also help examine the obvious, visible parts as well as the hidden and underlying parts of the system that are either affected by the system or affect the system. This chapter also repositions the role of the designers towards new roles of co-creators, facilitators, and collaborators, who simultaneously operate as designers of and for the system while being a part of the system. The chapter concludes by presenting a five-step design pathway that lets the designers frame and comprehend the system, envision and formulate the interventions, and lastly enable the transitions, supported by tools such as Gigamap and ZIP analysis.

Keywords Complex systems · Socio-technical systems · Wicked problems · Human-Centered design · Design methodology

V. Kant (✉)
Indian Institute of Technology Kanpur, Kanpur, India
e-mail: vkant@iitk.ac.in

V. Kant (ed.), *Human Centred Design for Systems*, Springer Series in Design and Innovation 68, https://doi.org/10.1007/978-3-032-30376-9_2

2.1 Introduction

The aim of this book and this chapter is to help human centered designers and provide them with new thinking tools to deal with systemic problems. It will also provide concepts and terms to help understand complex systems. The emphasis is on complexity and complex thought. This first chapter of the book will help you to understand how systems thinking offers a coherent intellectual field of study for designers and encourage you to apply these ideas in your own design projects. The key emphasis of this book will be on a practice-based approach to human-centered design for systems and how it can be achieved.

There are few things that we would want to consider. First and foremost, what are we learning out of this? This chapter will give you new thinking tools to deal with systemic problems. It will also provide concepts and terms to help understand complex systems. The emphasis is on complexity and complex thought. Further, this chapter will help you to understand how systems thinking offers a coherent intellectual field of study for designers and encourage you to apply these ideas in your own design projects. The key emphasis of this book will be on a practice-based approach to human-centered design for systems and how it can be achieved.

Specifically, we are looking at issues of human-centered design for systems. The underlying basis of this book is how can designers engage with systems. What are the new roles for designers? How the notion of design changes when systems thinking is brought in play? What kind of problems are specifically amenable for human-centered design when we look at systems and systems thinking? The chapters of this book will present human-centered design projects as examples to understand how designers can engage with complex systemic thoughts and use human-centered design tools to create value for the end users. The current chapter will present the concepts of systems design.

2.2 Systems Everywhere: Complex and Evolving

The first issue is to recognize that systems are everywhere. A number of existing systems that you encounter in your everyday life could include things like natural systems such as weather, weather as on a planetary basis, ecosystems, biospheres. These are very complex interrelated systems. In terms of man-made systems, human-made systems, there are information systems such as operating system, cloud-based services. Everyone uses mobile phones. Everyone uses technological apps, and you store your documents on the cloud. Therefore, cloud-based ecology as a system. Social systems such as languages, laws, organizations also exist. These are social in nature because of the multiplicity of people. The people interact and the coherence is brought about because of these entities' rules, laws and regulations.

Then there are hybrid systems. These are socio-technical systems such as health-care, air traffic systems. Sociotechnical systems mean people and technologies together are interacting with each other. These are people and technologies together in very complex cases of scenarios. For example, whenever you board a flight, right, the flight takes off, the flight lands. There are a whole group of people in the background making safety in these complex air traffic systems possible. That is also a part of socio-technical systems [1].

The important aspect that we have to remember is that there are systems everywhere. For example, there is transportation systems throughout the planet. Today, you can buy an equipment which has been made in let's say a different country such as Japan and you can be using it in India. The manufacturing of the chip may be taking place in Taiwan. The product design may have been done in Japan. The assembly may have been done in another country and then people are using it in a different country or globally, and all of this is supported through a robust financial system in place.

Hence, there are transportation systems, manufacturing systems, financial systems, information technology systems supporting all the global transactions. Comprehending, addressing, and offering inputs as designers to these global systems is a huge challenge.

Two very prominent words that are of interest to us are "complex" and "evolving". Complex and evolving are hallmarks of these systems that we see at a global level or even around us. Complexity means that there are a number of parts and these parts are connected to each other in non-trivial and myriad ways and they interact with each other in very interconnected manners [2]. In other words, these complex systems are interaction driven systems. Their identity as complex systems comes through because of its emphasis on the way interactions work in these system. The other dimension is evolution. These complex systems are partly self-correcting, and they can be sensitive to a number of disturbances. Over the course of time, these systems evolve. Therefore, complexity and disturbances together play a huge role in characterizing these modern day systems [2, 3].

2.3 When Is a Systems Approach Required? Wicked Problems, Uncertainty and Messes

One may wonder, when is a systems approach required? Whenever you have multiple perspectives on just what the situation is and how to deal with it. Also, if there is high level of ambiguity, uncertainty and complexity involved [3]. Whenever you are not guaranteed about what exact form of the outcome will be and the very situation is highly ambiguous, has multiple perspectives, a lot of stakeholders are involved, they have their own viewpoints. These kinds of situations require a more sensitive and robust systems approach.

Further, let's consider this situation. A previously applied fix has created problems elsewhere or moved beyond what was interested. Essentially, you applied a fix for a problem and it created another problem, eventually moving beyond what you wanted as a solution. While you thought it would be a good solution, it went beyond your original thought. These are areas which are very helpful, or these areas will benefit from using a systems approach. Also, when over a period of time you are settling for less than what you had aimed for, often because of the way things are set up, you end up tending to settle for lesser aspects. There a systems approach may be helpful. Additionally, in situations where the problem keeps reappearing, despite fixing it multiple times, then you may have to use systems approach. Here, you might be using the same fix repeatedly and systems approach may help you to arrive at a novel solution.

What kind of problems is the systems design process particularly suited for? Notice the word wicked problems or messes. Wicked problem was a description used by very famous systems theorists, Horst Rittel and Melvin Weber [4, 5]. Messes as a term was introduced by the systems theorist Russell Ackoff [6]. What they said is that see conventional ways of thinking about problems do not fit in with systems level problems.

They are just wicked problems that you cannot simply solve or in other words, they are messes. But the important question is why they are a mess. Let us recount a story you may have heard before about the blind monks examining an elephant. Figure 2.1 showcases a very famous painting by the Japanese artist Hanabusa and depicting a parable that you may have heard earlier. There are some blind monks engaging differently with the elephant. One of them goes and touches the ear of the elephant and describes the elephant like a fan. The other touches the trunk describes it like a snake. Another one touches the leg and describes it like a tree trunk. Similarly, all the other monks touched different parts of the elephant and described it differently. All the monks were unaware of the totality of the elephant, the broader aspects of the elephant, and only saw their own limited viewpoints.

They have their own perspectives on the problem. In order to solve the problems with multiple perspectives you will require the systems approach. These wicked problems have oftentimes low level of agreement on what the problem really is. Hence, just like the blind monks and the elephant, there is low level of agreement on problem definition as well as certainty about what to do about it. First, there is often a lack of clarity about what the problem is and second, there is a low level of certainty about how to respond to it. The third issue is that there is high degree of unpredictability due to the high number and diversity of stakeholders. All this matters and contributes to the formulation of a wicked problem [3, 4]. *The question for designer is what do we do with it and how do we address it? To reformulate the question—how can you, as a human centered designer, engage with systems?*

2.4 Designers of Systems, for Systems, in Systems,[…]

Primarily, we must recognize how designers can help in dealing with wicked problems. Design allows for understanding and representing. Understanding one aspect, representing the other. In other words, design helps you to understand and represent,

Fig. 2.1 Perspectives on understanding a system [7]

even recognize and formulate, various ways of approaching a problem as a systemic problem. Design also provides various visual tools, frameworks, ways of envisioning the problem and providing creative solutions. Human centered design as a discipline is well suited to support designers. They are well suited to support the activity of systems and dealing with these wicked problems.

The question that arises is—how, you as a designer, view the problem of systems design or how do you approach the problem when you are viewing the problem from the lens of systems design? As a designer, you must first grasp the complexity of the systems problem. The question then becomes—how do you deal with that complexity? The next thing you need to do as a designer is to design interventions that address the breadth and the depth of that complexity.

You, as a designer, are not only creating something for someone. There is multiplicity of roles that designers occupy in systems of various kinds. You, as a designer in systems, are designers of the systems, for the systems and often embedded within the systems. You are designers in systems.

In an information technology setup, you will be a designer working on the HMI which will influence the way services are sorted out or you may just be focusing on services in an air traffic system. The key idea is that for designers, there is not one role that they have in dealing with systems and they should recognize that they must be adaptable and flexible in dealing with systems.

Let's look at the term designer. The traditional views of design that we have are designers as creators of products, or designer as creator of messages in terms of communications, communicating something, or designers as creator of interactions, interaction designers.

These standard roles of a designer as a creator are changing. Reason being, when you are dealing with systems and systems thought, you are dealing with multiple roles, multiple perspectives, multiple ways in which people sort out their problems. As a result, you are now acquiring a new role as that of a co-creator, as that of a facilitator in which you empower people to come up with their own solutions. Here, the role of designer is not just the role of a provider of services and products but also of a participant. Design as co-creation. Design as facilitation. Design as collaboration. Design as enabling. Design as empowering.

All these aspects exist in systems and systems thought and its applicability in human centered design for systems. In other words, as designers, our basic focus should be on asking how we can create value and meaning for people who use these systems and are involved within these systems. This holistic perspective is very important when we are thinking about designing for people within these systems and designing for people who will use these systems.

A brief understanding of the background about how systems thinking has come up in the past century will further build on this discussion. It will also help us in formulating newer responses in how we can go ahead and develop systems and systems thought or how we can engage with designers or how we can engage design with systems thinking.

2.5 Background of Systems Thinking for Human-Centered Design

In the past century, there have been multiple viewpoints and multiple sources on how systems and systems thought have evolved. Systems and systems thought have evolved from early 1900s coming in from ideas from biology, mathematics, physiology, philosophy and sociology [8–12]. They contributed to what was known as a new approach towards thinking about science. The researchers during this time started to recognize, especially in areas like biology, that a reductive approach of reducing phenomena to base level variables is not enough. Eventually, they started to ask what kind of newer science would be required to understand natural phenomena in the best manner possible. This was one approach that came through and all these ideas started coming together and they formed what is now known as a general systems theory [13]. The focus of this thread of general systems theory was how can we incorporate these newer phenomena or how can we see phenomena in real life which such that we can move beyond a reductive approach towards a more holistic approach.

The second viewpoint also came from cybernetics and second order cybernetics. Cybernetics was a field that started in post-World War II by a number of researchers, prolific researchers, and they came up with their ideas about how to think about control in humans, machines, animals [8, 11, 13]. In other words, they started looking at a general viewpoint, a general discipline or a general field of thought where they can use first order and second order cybernetics to think about control and communication in various kinds of entities, humans or animals or machines. All of these started coming together and along with this there have been system sciences arising from various interdisciplinary movements in science.

Systems engineering has also been coming through. There are also applied system studies. All of them together started to form what is now known as a plethora of systems approaches in various fields of study ranging from engineering to cybernetics to management and as well as later they have now also been started to be used in design.

In design, the usage of system theory in design is not new. In design schools in the 1960s there was a very famous school of design called the Ulm school of design from Germany which has left its impact on design education and modern design thought. In that school, designers at that time were taught about systems theories and cybernetics, i.e. in the 1950s and 60s [3, 12].

The question that arises is—how can you use systems science and connect it to systems design and use it for formulating design and design approaches. This is where systems design started to get connected. Modern systems design as we know has come out from this mode of enquiry from the Ulm school of design as well as other design approaches started acquiring it [12].

As of today, there are multiple approaches to systems design currently being practiced. Two very prominent approaches have been systemic design and systems-oriented design. Both of these falls under the very broad category of relational systems design [14]. The focus of relational systems design has been on terms such as looking at relations between entities and emphasizing or capitalizing on these relations to formulate the next generation of systems thought as well as using it in design practice.

The rest of the book will take the notion of relational systems design, opens up the elements that are required for systems design through this approach. As well as use it for design practice to demonstrate how design projects can be accomplished using this approach. Now, how do we place systems design, as related to other design disciplines.

So, some of you coming from the design disciplines, may be trained in communication design or even industrial and product design right or interaction design. All these are more popular choices in design or popular branches of design that you may have heard about, but one way of envisioning it is through Richard Buchanan's way of thinking about orders of design. Buchanan addresses thinking about design more broadly [4] and his thoughts have been developed by Golsby Smith [15] about how we can think about systems design as a higher order of design (Fig. 2.2). Think of the number of nested approaches that you are seeing in these in on the screen. For example, you are seeing communication design, and which is nested in industrial design. So, the idea is that industrial design also requires communication design. For example, in packaging of products.

In turn, Interaction design would also involve industrial design and or the products of industrial design and communication design. Suppose you have to design a mobile

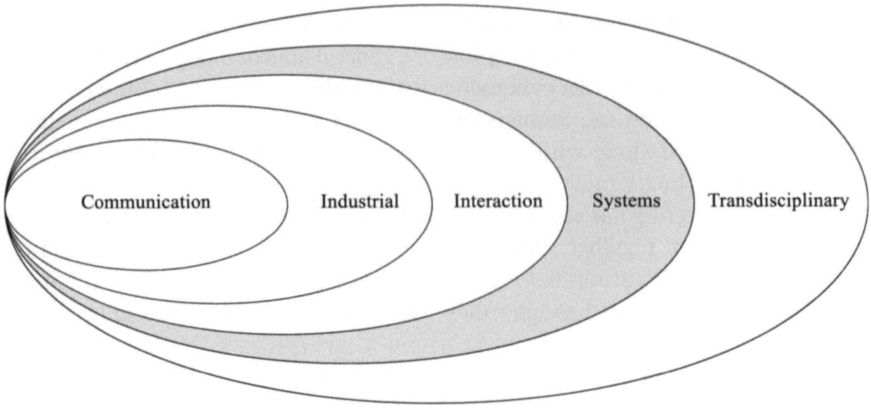

Fig. 2.2 Locating systems design within the design disciplines [4, 15]

phone, then you are thinking about how the phone will be held, the physical form factor, the dimensions of the phone how you will use it, you will also think about what are the visual communication messages that you will focus on it.

We can think of systems design as a higher order of design, where to do systems design, you would require a strong foundation in the traditional design disciplines of communication industrial and interaction. But at the same time system design is not reduced to these various aspects of design. Systems design would involve a development of all these various branches of design to go ahead and come up with a broader level of thought process, broader level of envisioning. How can these various outputs of design disciplines come together to solve wicked problems [4]?

Systems design is not the final aspect of design disciplines. In general, designers and design researchers have been recognizing that there are transdisciplinary dimensions of design making systems design a nested dimension. There are more higher dimensions to design in the form of transdisciplinary design [15].

Systems design is nested between these broader realms of design as well as lesser orders of design [1, 3]. This enables us to recognize that systems design can be very meaningfully engaged with other broader design disciplines.

2.6 Vocabulary to Engage Systems as a Human Centered Designer

How do we address systems? In other words, we need some vocabulary to address systems. First and foremost, what is a system? Typically, when we think about systems, we think about how a system consists of different parts, different entities that are put together to form a coherent whole. Which means the parts have certain kind of interrelationships, the parts interact with each other, and they interact with each other in non-trivial ways. The interaction holds value. This would constitute a system at a holistic level.

Let us see an example. Body organs can be taken together to form a digestive system. So, the organ and its behavior may be very specific, but all the behaviors are taken together and fit together in form of a broader level digestive system.

Another example, in terms of technical components, would be parts of a car put together to form a machine. Russell Ackoff was a very famous systems theorist who gave the example of his car to demonstrate why you have to think in terms of systems [6]. Say you have taken the best elements or best parts of different cars throughout the world—the wheels of a very famous brand such as a Ferrari, the body shape of a very good vehicle such as an Audi, or the transmission of a very famous brand such as Mercedes. Now you cannot simply cobble these parts together to create a car and expect that the car will perform at the best possible way because it has the best parts. The important idea is to understand that it is not only at the level of parts that you need to think, but at a holistic level how a system is formed.

Russell Ackoff defines a system as a set of two or more elements that satisfy the following conditions. The first condition is that the behavior of each element has an effect on the behavior of the whole. Second, the behavior of the elements and their effect on the whole are interdependent. However, subgroups of elements are formed such that each has an effect on behavior of the whole and none has an effect independent on it. Notice how subgroups are formed, how individual elements are there and how all of them interact with each other [6].

Along with this definition a more picturesque way of thinking about systems is in terms of metaphors. Metaphors are parts of speech, where you refer to something else that has similar characteristics. So, metaphors imply one thing that is another. A very common metaphor that systems theorists use are icebergs. An iceberg is a block of ice floating on the sea. Whatever you see of the iceberg is a small part which shows above the surface of the sea, the rest of the iceberg is below the surface [2, 3]. What it implies is that in any system, there are very limited aspects of system that you are really seeing. The rest of the system is hidden.

Another metaphorical way of looking at the systems is that of a tree, which has multiple branches and roots. Like a tree, a system spreads its parts, connected to each other. Interconnected relationships or webs is one more way of thinking about the systems [1, 14]. Notice how the interconnections hold up the web together, in the way it is built up right.

Icebergs, trees, and webs are different ways of thinking about systems in a coherent manner. These metaphorical ways of thinking about the systems gives designers a more a broader way of thinking about systems rather than very specific definitions.

As a designer if you are engaging with a system, the kind of questions you can ask as a designer can be metaphorically shaped. For example, if your think of the system metaphorically as an iceberg, then you can ask questions like—what are the visible variables? What are the visible connections? What are the invisible connections and variables? You can ask what the deep-rooted myths and the belief systems that people have within the system. These are a set of questions which enable you to delve into understanding the system in greater detail.

Further, elements of system include a boundary. A boundary is an element which separates the system from its environment. Oftentimes, these are not sharply defined but still you have to recognize, that where there is a system there is an environment. An environment, in many cases, plays an important role in how you are engaging with the system. Additionally, have to think of the system in terms of coherence mechanisms. This means, the parts of the system are interacting with each other so that there is a certain coherence that is given to the system. The other aspects are that you have to think about the system are in terms of observable interactions and non-observable interactions.

Observable interactions include what is straightforward and what you can see in the functioning of the system. They are very obvious, how the system interacts with its environments, inputs, outputs, things which are clearly visible. But then there are non-observable interactions. You have to infer that there are certain processes in play. In other words, there are certain ways of thinking. These are challenges or these are elements or the basic vocabulary you have to consider while you are engaging with systems.

The first step in systems comprehension is addressing the concepts of parts and wholes. Christian Von Ehrenfels, an Austrian philosopher, very early in the last century highlighted holistic phenomenon in our everyday life. The example he gave is that when you are listening to a piano, you hear the melody and you hear the melody as a whole, not at the level of individual notes. After him these Ehrenfels qualities are named. Ehrenfels qualities basically mean that these are holistic qualities by which people understand and comprehend phenomena. The key idea is that there is the need to recognize that there are many phenomena that are at the level of the wholes, not simply reducible at the level of its parts.

Another way of thinking about it is in terms of complexity. In other words, the parts define the whole and the wholes define the parts. This circular logic where the level of parts defines the level of wholes and in turn, the level of wholes define the level of parts, making the system complex [3]. This interaction of parts and wholes is a very important dimension when you are thinking about systems and engaging with systems.

We have been talking about parts; we have been talking about wholes and now we will talk about what kinds of relationships exist between parts and between wholes. In order to understand the notion of relationships, let's take the very famous example of these cartoon characters Tom and Jerry. I am quite certain many of you, as kids, have watched Tom and Jerry shows that have been broadcast a lot on television. In the show, there the cat, Tom, is one entity, and the mouse, Jerry, is another entity. They are sometimes portrayed as enemies and at times Tom chases Jerry. At the same time, in many cases, Tom and Jerry act out as friends. So, their relationship of a cat versus a mouse is not always static. It is a very dynamic relationship that evolves over time. In addition, there are other characters also, the bird, the dog, the other stray cats and then there is Jerry's nephew (the small grey mouse). All these other characters also interact with the two main characters of Tom and Jerry. In Fig. 2.3, you will notice in the center, Tom has certain interrelationships with Jerry and Jerry

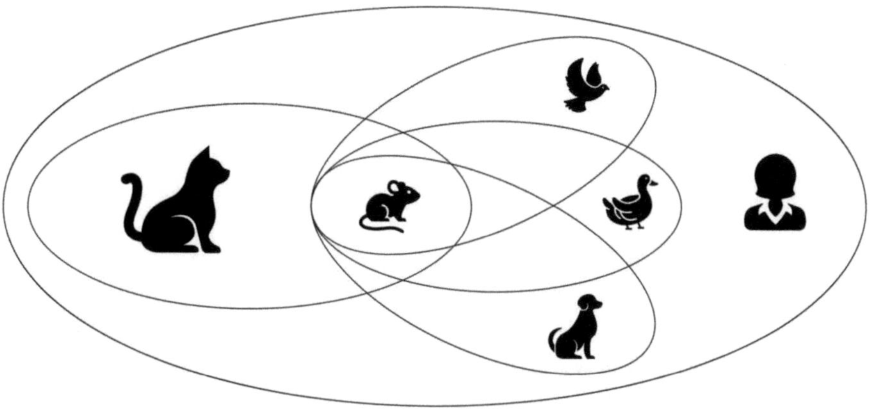

Fig. 2.3 Relationships between entities of a system using the example of interactions between cats, mice, people and other stakeholders

has certain interrelationship with other characters. All of these affect the relationships that they have with each other.

This is why relationships play a very significant role. Relationships are important for understanding systems and characterizing systems. In turn, what we are now seeing is that relationships are also defining these characters in some way or the other. Relationships are very crucial in systems and human centered designs for systems, and these relationships show how various entities are influenced by each other.

We have now completed a number of elements in the vocabulary of systems, while being a human centered designer.

2.7 Relation Between Systems Thinking and Design Thinking

Building on this relationship and we can look at the relationship between systems thinking and design thinking, a little bit more closely. In Fig. 2.4, you will see reflection on the top left part. Systems thinking, design thinking, and futures thinking are more reflective in their approach. The lower part involves a more action-based approach that of systems practice, design practice, and futures practice. Another aspect that you will observe is that we have systems on the left, design at the center and futures on the right and there are various interrelations between them [3].

Let us say you are completely on the left side, and you are interacting with more systems thinking and systems practice. Then you will find yourself to be on the side of engineers and managers. A lot of engineers and managers use systems thinking and systems practice. On the other hand, on at the center you see designers who are constantly engaging with design thinking and design practice. Our focus is on

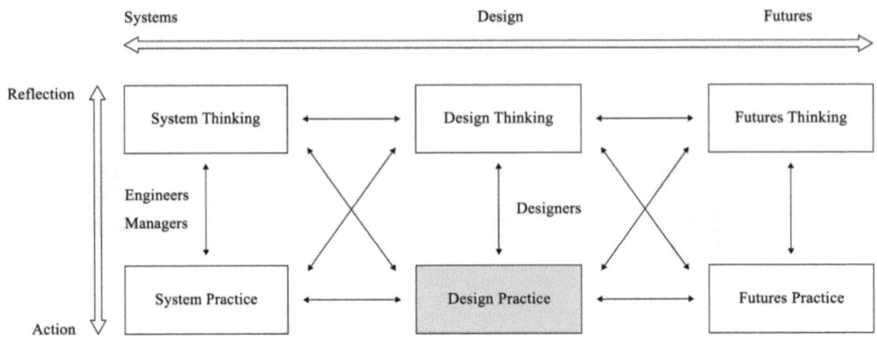

Fig. 2.4 Relationship between systems thinking, design thinking and Futures thinking. expanded from [3]

these interrelationships in terms of design practice where design practice is informed by design theory and design thought. It is also informed by systems thinking and informed by systems practice.

Hence, as a human centered designer your focus will be on design practice, but your thought process and your design practice will be heavily influenced by systems thought, design thought and systems practice. All of these are going on in light of the major changes in terms of the futures thought.

Having explored different ways in which we interact and engage with systems, it is important to recognize certain keywords that will orient us towards systems. So far, we have asked how we engage with systems, engage with people within the systems and how does the role of a designer change in such contexts. The question now changes to how a designer should orient themselves towards a systems project. When you are approaching your projects, you should keep a few keywords in mind such as participation, facilitation, engagement, co-creation, design practice, design thinking, systems practice and systems thinking. All these keywords orient us towards how we should be a human centered designer in systems [1, 2]. You will see these aspects become more prominent in the other projects that we take up in the subsequent chapters of this book.

The question that follows is—if you were given a design problem, a systems design problem, how will you address the systems design problem as a designer? What kind of a design approach will you use? What kind of a design methodology you would use to address this issue?

2.8 Pathways for Engaging Systems a Human Centred Designer

The first aspect you should be thinking about is the pathway for engaging systems as a designer. The key idea is to scope the extent of your problem space. Notice that you have to in some way or the other create a boundary for your system which scopes down your problem. In other words, what is the system or part of the system that you are looking at?

Then you will always recognize that there is an environment which is affecting that system. This scoping allows you to understand the extent of the problem space. The second step you should be taking is to discern a set of distinctions and common patterns. What kind of relationships exist in that system, and you can describe the functioning of the system through visual methods that designers use.

A lot of mapping methods, diagramming methods can be used to make the non-obvious, the latent functions, the ways in which the system functions, to be made apparent. Once we have diagrammed or mapped out the system in human centric terms, the emphasis on human centric terms, we see there can be many representations to systems. However, some very important representations of systems involve representing the systems in a manner that is useful for the designer to intervene into the system, and this aspect will be seen by examples that we see through the rest of the book.

Our focus will be to identify how we can engage with these systems, what are the possible ways in which we can represent these systems, and what are the possible points at which we can intervene into the system. In order to intervene into the system, we will look at the system through something known as the ZIP analysis [3], i.e., zooming into areas that require more depth, understanding the problems, pain points, and potential points, and finding the intervention points, where the solutions for P-points can be found. Once we identify the intervention points, we can design interventions and make sense of the impact of those interventions on the rest of the system.

This is a five-step method by which you can frame the system, comprehend the system, envision the interventions, formulate the interventions and then enable the transition [2, 3].

The first step, framing the system, involves understanding the context, defining the specific requirements, finding the constraints and then defining the scope of your problem. It also entails outlining the boundaries of the system (Fig. 2.5).

The second step, comprehending the system, involves identifying the key stakeholders, mapping them in different structures, mapping the connections withing the system, and recognizing the long-term trends, structures, practices, and so on. All the possible dimensions of the system are captured using a diagramming method called the Gigamap (Fig. 2.6).

A Gigamap is a collection of various maps or various representations, perspectives on a system and how they are interconnected to each other [3]. Remember the blind monks and the elephant [7]? Metaphorically, the system is the elephant over

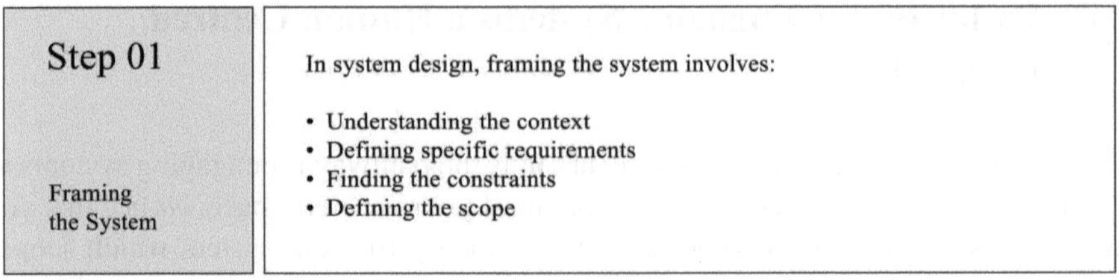

Fig. 2.5 Step 1 of the human centred pathway for engaging the system—framing the system [2, 3]

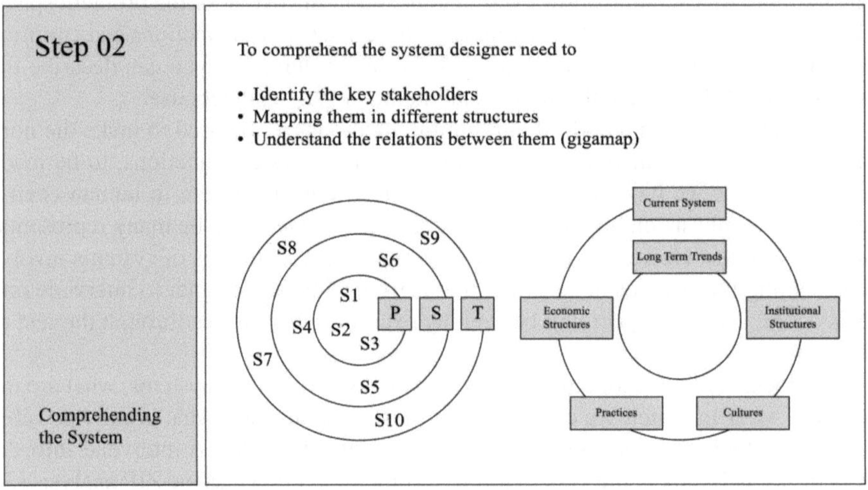

Fig. 2.6 Step 2 of the human centred pathway for engaging the system—Comprehending the system [2, 3]

here, and we, as designers, are the blind monks trying to understand various parts and dimensions of the system. Through this different perspectival approach, we are putting all of it together in this giant mapping scheme which is called the Gigamap.

Once we have a Gigamap, the third step is to identify the intervention points. In creating this diagrammatic representation, we often find that there are certain areas where the system interventions would have a greater impact. We zoom in on these points in depth (Fig. 2.7).

After identifying the zoom points, we highlight the potential pain points, points of problem or small regions of problems. Then we innovate by removing the bottlenecks or bringing in new relationships. We bring in ideas, interventions, innovations to make the life of the people involved in the systems easier. We can do this by marking the Gigamap with labels of Z, P and I (Fig. 2.8).

The zoom points can be marked with Z1, Z2…, the potential pain points can be marked with P1, P2…, and the possible intervention points as I1, I2…. We can mark multiple points to get clarity on the problem. Based on our clarity, we introduce newer

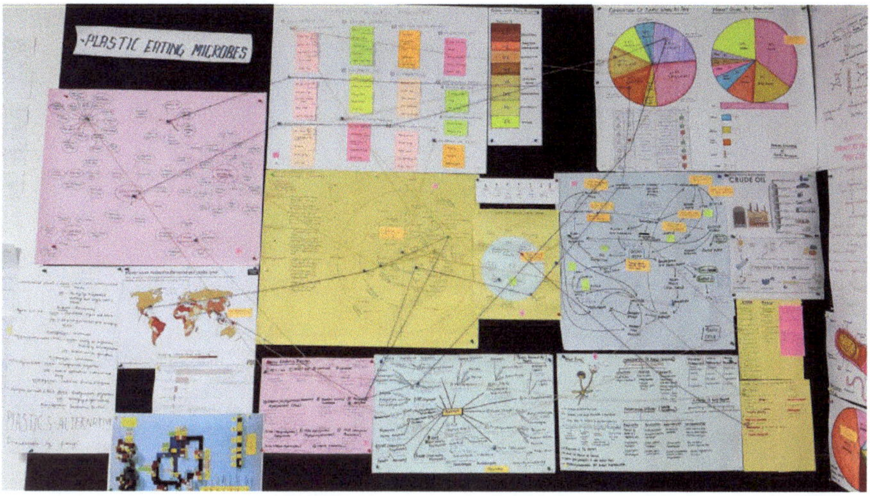

Fig. 2.7 Sample giga map—ecosystem of plastic-eating microbes from Chap. 8 of this book

Step 03 Envisioning Interventions	Identify Intervention Points (ZIP Analysis - I): • Zoom ▪ Points where a systems intervention would have a greater impact on the whole system • Innovation (idea, intervention, innovation) ▪ Innovate by removing bottlenecks, or bringing new relationships • Potential (problem or 'painpoint') ▪ Points of problem or small regions of problems

Fig. 2.8 Step 3 of the human centred pathway for engaging the system—Envisioning Interventions [2, 3]

intervention areas, newer innovation areas that will have potential outcomes. Once the ZIP analysis is complete, we can also understand the system in terms of abstract trends and characteristics that would change when we are describing it (Fig. 2.9).

When we are formulating the interventions, fourth step of the process, we not only focus on the specific aspects of the problem but also how it will impact the whole system around it. We also consider the impact in different layers, recognizing the key factors or classification withing the system. For instance, in case of a technological system, one might think about the components which is at the level of work, we can also think about it in terms of assembly which is at the level of organizations, or the factory which is at the level of regulations. The same system can be understood in terms of different levels or layers. With a focus on classification, factors, and abstract trends, we can to formulate an intervention (Fig. 2.10).

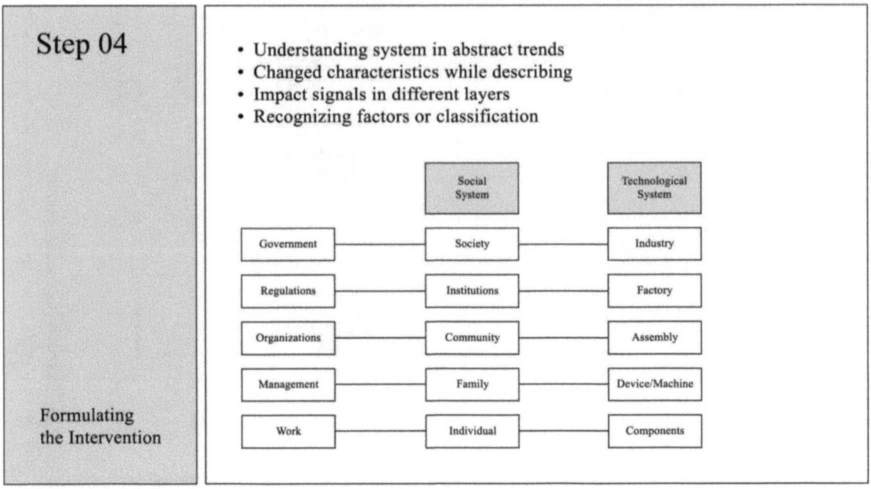

Fig. 2.9 Step 4 of the human centred pathway for engaging the system—formulating the intervention [2, 3]

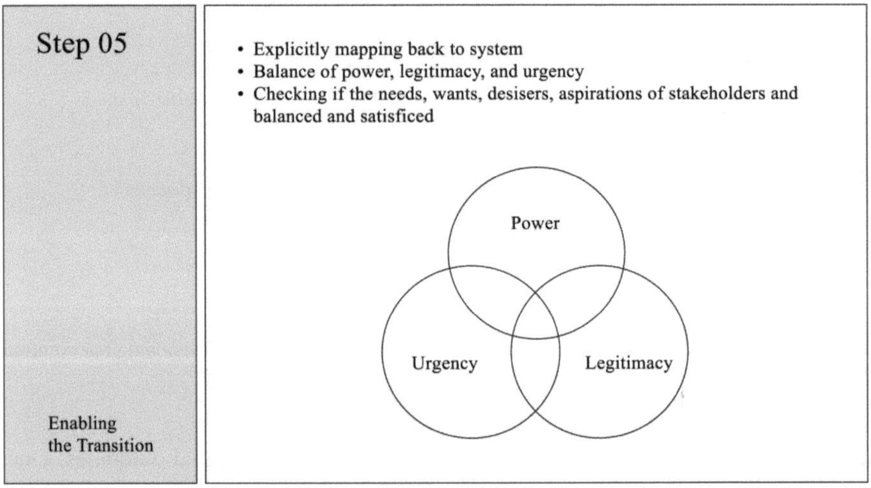

Fig. 2.10 Step 5 of the human centred pathway for engaging the system—enabling the transition [2, 3]

Finally, once we have formulated the intervention, we then explicitly map it back to on the system and assess its potential impact and challenges to decide if we were to bring about this change. This will involve understanding how introducing new ways of working might affect the balance of power, legitimacy and urgency within the system as well as the needs, wants or desires of the stakeholders. This will raise critical questions such as are all the stakeholders satisfied? Are the needs of some stakeholders getting affected or balanced? These broader questions will reveal the implications of the intervention.

At this stage, all the aspects that we have studied in terms of vocabulary of systems—uncertainty, wicked problems, metaphors, design process, that are required to address systems problems need to be explained in greater detail, with a focus on design practice. This is because the focus of this book is on using design practice and how we can use these systemic terms and engage systems theory for design practice. A better way to do it is through demonstration.

To demonstrate this practice, the following chapters present nine different examples across different sectors, ranging from plastic-eating microbes to disaster ethics to the use of millets. Each example applies our systems design approach to come up with an intervention that will be meaningful to the people within the system as well as the system as a whole. We practice what we preach!

References

1. Manzini E (2015) Design, when everybody designs: an introduction to design for social innovation. MIT Press
2. Jones P, Van Ael K (2022) Design journeys through complex systems: practice tools for systemic design. BIS Publishers, Netherlands
3. Sevaldson B (2022) Designing complexity: the methodology and practice of systems oriented design. Common Ground Research Networks
4. Buchanan R (1992) Wicked problems in design thinking. Des Issues 8:5. https://doi.org/10.2307/1511637
5. Rittel HWJ, Webber MM (1973) Dilemmas in a general theory of planning. Policy Sci 4:155–169. https://doi.org/10.5040/9781474282932.0015
6. Ackoff RL (1994) Systems thinking and thinking systems. Syst Dyn Rev 10(2–3):175–188
7. Itcho H (2001) Blind monks examining an elephant. From Library of Congress, accessed on 8 April, 2026. Accessed from https://www.loc.gov/pictures/item/2004666374/
8. Ramage M, Shipp K (2009) Systems thinkers. Springer, London, pp I–VII
9. Midgley G (ed) (2003) Systems thinking. Sage, London
10. Midgley G (2000) Systemic intervention. In: Systemic intervention: philosophy, methodology, and practice. Springer, Boston, MA, pp 113–133
11. Checkland P (1981) Systems thinking, systems practice. Wiley, Chichester
12. Trivedi K (ed) (1989) Readings from Ulm: selected articles from the journal of HfG Ulm. IDC Press, India
13. Von Bertalanffy L (1973) General system theory: foundations, development, applications. George Braziller, USA
14. Escobar A (2018) Designs for the pluriverse: radical interdependence, autonomy, and the making of worlds. Duke University Press
15. Golsby-Smith T (1996) Fourth order design: a practical perspective. Des Issues 12(1):5. https://doi.org/10.2307/1511742

Chapter 3
Disaster and Ethics: Aftershock of Choices

Kashika Chawla, Tanishka Chavan, and Abhigyan Deka

Abstract Ethics form the foundation of any humane and orderly society. Conversely, earthquakes are among the most destructive natural disasters that have the potential to cause severe loss of life, and widespread socioeconomic disruptions. The unpredictability of such an event requires prompt decision-making, which may adversely affect the affected communities and cause additional stress and moral burden on those responsible for making these choices. This study bridges that gap by examining the ethical challenges in earthquake relief efforts. The exploration of this topic focuses on three key aspects: disasters, ethics, and decision-making. These aspects were visualised based on the metaphor of 'Jenga blocks', where each block plays a crucial role in making the structure stable. Similarly, in this project, these three aspects play an important role in stabilising the system of ethics in earthquake response, wherein the absence of any one aspect could easily lead to a cascading failure of the entire system. The system was visualised using a Gigamap, which helped understand the ethical dilemmas that arise at different stages of an earthquake, and investigate the relationships between categories that may initially seem unrelated. The proposed intervention reframes ethics as a proactive tool rather than a reactive concern, which would encourage individuals and institutions to integrate ethical thinking into everyday preparedness. This project emphasises that ethical awareness is not limited to a theoretical notion but is an important component of effective disaster management. It also highlights how moral responsibility, empathy, and informed decision-making can significantly improve outcomes for affected communities.

Keywords Disaster · Ethics · Disaster response · Disaster management · Awareness

---

K. Chawla (✉) · T. Chavan · A. Deka
Indian Institute of Technology Kanpur, Kanpur, India
e-mail: kashikac24@iitk.ac.in

V. Kant (ed.), *Human Centred Design for Systems*, Springer Series in Design and Innovation 68, https://doi.org/10.1007/978-3-032-30376-9_3

3.1 Introduction

Disasters are sudden, disruptive events that cause significant harm to human life, property, the environment, or essential systems, often overwhelming the capacity of affected communities to respond. These events may arise from natural hazards such as earthquakes, floods, or cyclones, or from human-made causes such as industrial accidents, armed conflict, or environmental degradation. While a hazard is a potential threat, it becomes a disaster only when it leads to widespread damage and human suffering. Earthquakes are among the most devastating natural disasters, causing widespread destruction and loss of life. 59% of India's landmass is prone to earthquakes. The total area of the country is divided into four seismic zones, where Zone V is the most active region and Zone II being the least active region [1].

Ethics are the moral principles that guide decision-making and govern human behaviour. They involve fairness, responsibility, and respect for others in decision-making, ensuring actions align with the values of justice and integrity. Decision-making refers to the process of evaluating alternatives and selecting a course of action. It involves making choices among different options and can be influenced by factors such as knowledge, speed, accuracy, and uncertainty [2]. Ethical decision-making in disasters is critical. Human lives, required resources, and recovery of the affected individuals and communities depend on it. When disasters such as an earthquake strike, the urgency of response often leads to ethical dilemmas faced both by the affected and the responders.

3.2 Framing the System

Ethics in disasters refers to the moral principles and frameworks that guide decision-making in high-stress, high-stakes situations. Disasters, whether natural or human-made, often demand urgent choices about resource allocation, prioritisation of aid, and interventions that can significantly impact human lives. In these moments, ethical decision-making becomes critical to ensure fairness, accountability, transparency, and respect for human dignity. These decisions are rarely simple and often involve navigating complex dilemmas where competing values, interests, and constraints must be carefully balanced.

Ethical considerations in disaster response are rooted in a range of sources, such as philosophical thought, cultural and religious traditions, professional codes of conduct, legal frameworks, and personal moral upbringing. Together, these shape the values that guide both individuals and institutions in crises. Without a strong ethical foundation, disaster responses risk becoming arbitrary, discriminatory, or even harmful, undermining trust, justice, and long-term recovery. While most existing disaster response models focus on logistics, coordination, and technical efficiency, ethical dimensions are often sidelined. Yet without deliberate attention to ethical principles, even the most efficient response can fall short in addressing human needs with compassion, equity, and integrity [3].

Ethics in the context of earthquakes is a part of a larger of a socio-technical system, where human values and behaviour interact with technical systems, infrastructure, and institutions. Due to the vastness and complexity of this dynamic system, it must be analysed both at the level of its parts, by examining specific people and situations, and holistically, by looking at the system as a whole across the country.

3.3 Comprehending the System

To establish the fundamental structure and define the boundaries for this system, stakeholders engaged throughout an earthquake were mapped (Fig. 3.1). All the background material used for analysis were open access reports, content from newspapers and news channels, and other sources on the internet. They have been collectively observed, comprehended, and used for this project. Primary stakeholders were considered to be those directly present during or immediately after the earthquake, such as victims and first responders. Secondary stakeholders were those who provide essential support to the primary group, including NGOs and relief organisations. Tertiary stakeholders, while not directly involved in the immediate response, tend to play a significant role in shaping long-term recovery and resilience. This includes policymakers, research institutions, and other entities that influence systemic decisions. Categorising stakeholders in this way allowed us to better understand the layers of involvement and the ethical dimensions each group may encounter [4].

Upon identifying the stakeholders involved, the key characteristics influencing stakeholder behaviour were analysed by using the stakeholder dimensions tool. It also led to the identification of the constraints and priorities that can either be based on their standing or their associations. For example, a paramedic would have different constraints as opposed to a police officer. A Samaritan volunteering during relief would prioritise providing aid to people in need, while a volunteer associated with an organisation would prioritise according to their training and protocol. The next layer of analysis included mapping the stakeholder roles that evolve across the different phases of a disaster, spanning from preparedness to response and recovery. These stakeholder actions are dependent on their seat of power as well as the sector they come from. Some stakeholders are associated with all three phases of the disaster, while others only participate in one or two phases. For example, government doctors during the preparedness phase take part in setting the policies and protocols for medical aid and allocating the resources, while in the response phase, their role is to provide aid for the victims. In the recovery phase, their focus is on long-term recovery of the victims as well as contributing to the updation of policies and protocols [5].

The next step was exploring what and how ethics relate to earthquakes. Ethics are the moral principles that guide decision-making and govern human behaviour. In disaster relief, ethics aren't just about morality; it's about making fair, responsible, and transparent choices when time is limited and stakes are high. Ethical principles

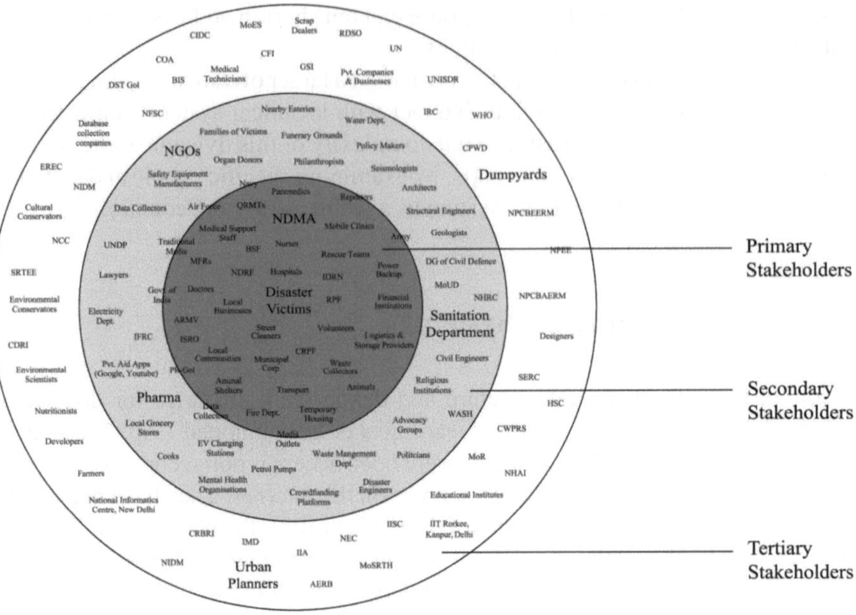

Fig. 3.1 Stakeholder mapping

such as solidarity, responsibility, non-discrimination, and cooperation serve as the foundation for disaster response. However, their application changes across different stages of a disaster. Before a disaster, the focus is on prevention through education, awareness, access to justice, and protective measures for vulnerable groups. During a disaster, ethical action involves providing humanitarian aid, respecting human dignity, and supporting those most at risk. After the disaster, ethics guide long-term recovery efforts by promoting resilience and protecting social, economic, and political rights. Understanding these principles across all phases helps ensure disaster response is not only effective, but also just and humane [3].

When the stakes are high and ethical priorities shift across phases, tensions can arise. Conflicts may emerge when different groups, each guided by distinct ethical principles, work toward a common goal. Here, conflicts may arise when military forces get involved in relief operations. Humanitarian efforts emphasise impartiality, neutrality, and community involvement, whereas military operations balance humanitarian needs with strategic and security goals [6]. While they bring efficiency, their presence may conflict with humanitarian neutrality. When ethics collide between different actors or guiding principles, it puts immense pressure on how decisions are made in the moment. In disasters, these decisions are time sensitive. Hence, it was crucial to understand the psychology behind crisis decision-making, as the tools used to respond ethically are cognitive, along with being institutional or procedural. Decision-making in such high-pressure contexts is deeply influenced by human psychology and behaviour. The concept of dual-process thinking explains the two ways in which the human brain processes information and makes decisions [7]. System 1 is fast, automatic, and intuitive, allowing for quick responses, but it can be prone to biases and errors. System 2 is slower, more deliberate, and logical, which is better suited for complex reasoning, but often tends to be time-consuming in high-pressure situations like disaster response, where time is limited. These processes operate together, but the balance between them shifts depending on the urgency and complexity of the decision at hand. This is where decision-making heuristics come into play. Heuristics are mental shortcuts that allow people to make judgments quickly and efficiently, especially when operating under time constraints and limited information. However, while useful, heuristics can also lead to systematic errors. For instance, the availability heuristic might cause a responder to overestimate the likelihood of a certain event based on recent media coverage, while the representativeness heuristic could lead to misjudging a new situation based on its similarity to a past event [7].

Any given system has individuals, groups, or organisations that possess significant power and vested interest, and whose support or opposition could have a significant impact on the system. The primary stakeholders were initially narrowed down, and then the key actors in the system were identified by analysing the stakeholder actions and variables together. These key actors were then mapped out on the knowledge-power opposite diagram, wherein the key actors holding greater power or knowledge essentially indicate their position or hierarchy within the system. Next, the limitations faced by victims and rescuers that may hinder earthquake relief were identified. These were categorised into three areas: unique limitations for victims, for responders, and shared obstacles between them. These limitations are illustrated in Fig. 3.2. For example, connectivity issues and language barriers are common limitations that would be faced by both the victims and the rescuers. These limitations would further add hurdles such as miscommunication, delayed relief efforts and many more. These hurdles are both influenced by and have an impact on the ethical facet of this system.

Apart from groups with opposing sets of ethical principles, ethical challenges may arise at an individual level as well. These challenges would either add to each other or contradict. For example, a paramedic may face an ethical challenge due to a language barrier. On the other hand, a responder may face an ethical dilemma when their values conflict with the requirements of their employing organisation. If both groups and individuals face certain ethical dilemmas in the event of an earthquake, it would have an impact on the decisions taken by them. Another aspect of this would be whether the decisions taken are ethical or unethical. In the case of ethical decisions taken, the result has a possibility of being negative, but it would be unintended and unforeseen. However, in the case of unethical decisions taken, it is highly probable that, despite the outcome being positive, it could result in some serious consequences. These consequences would not be limited to a short period but could easily have long-term complications. This would also create a loop of consequences that may keep affecting each other. The loop consequences are illustrated in Fig. 3.3. Continuing with the case of the previously mentioned responder, assuming that he has faced an ethical dilemma where his values conflict with the organisational needs. He decides

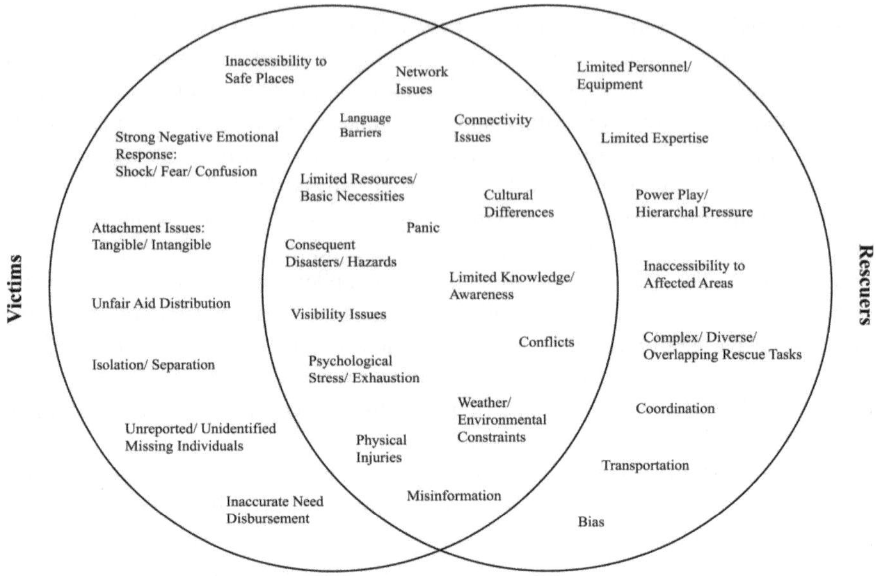

Fig. 3.2 Limitations that Hinder earthquake relief

to prioritise his values and takes an unethical decision. This decision leads to a short-term consequence, such as a loss of public trust. In the long run, the loss of public trust would cause a loss of reputation and credibility not only to the individual but to the designated organisation as well.

This is not limited to how people think; it also extends to what people feel. Psychological responses encompass the mental, emotional, and behavioural reactions individuals exhibit in response to internal or external stimuli or situations. They evolve through different disaster phases from the Impact phase to the Reconstruction Phase.

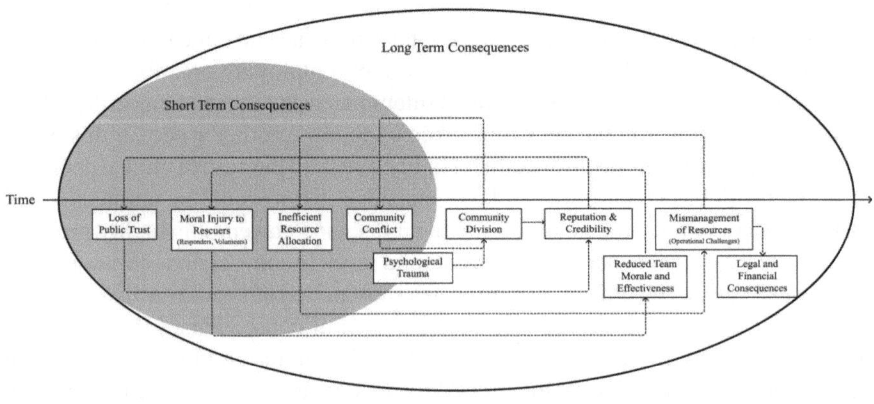

Fig. 3.3 Consequences of unethical decision-making

Understanding these phases helps provide the necessary support, both immediate and long-term, during recovery. Figure 3.4 explains the affected communities' and individuals' psychological responses across the different disaster phases [8]. Taking an example an individual staying in seismic zone four. Despite receiving earthquake warnings, if the individual is unable to get his family to safety before the earthquake hits, then he would feel guilty and blame himself. When the earthquake hits, he might have heightened emotions of confusion, shock, self-preservation, and protecting his family. In the week following that, he would face low productivity but would like to be constantly active. In the following 2–8 weeks, he would face a range of emotions, like optimism—a hope that everything will return to normal. However, he would also realise that there are limitations to the disaster assistance he can avail. In the next phase, which could span up to 36 months, he would face long-term exhaustion, and his previous optimism would turn to discouragement. In the last phase, he would face continued grief but would start adjusting to his new normal state of life.

To better understand how these psychological patterns play out in real contexts, scenarios for different primary stakeholders involved in an earthquake were developed. These scenarios helped explore how individuals might respond when faced with internal pressures, like stress or personal biases, and external constraints, such as limited resources or unclear information. Through this, different kinds of ethical questions that could emerge, and how heuristics might shape the decisions they make in high-stakes situations, were identified. For example, consider a community volunteer who reaches a relief camp after an earthquake has hit. He needs to cater to three victims. The first victim is a 35-year-old moderately injured male with a bleeding

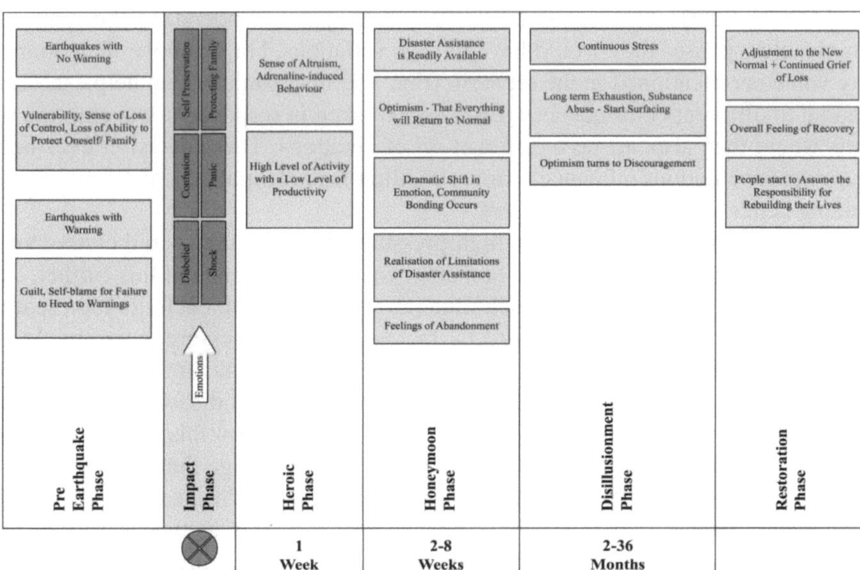

Fig. 3.4 Consequences of unethical decision-making [8]

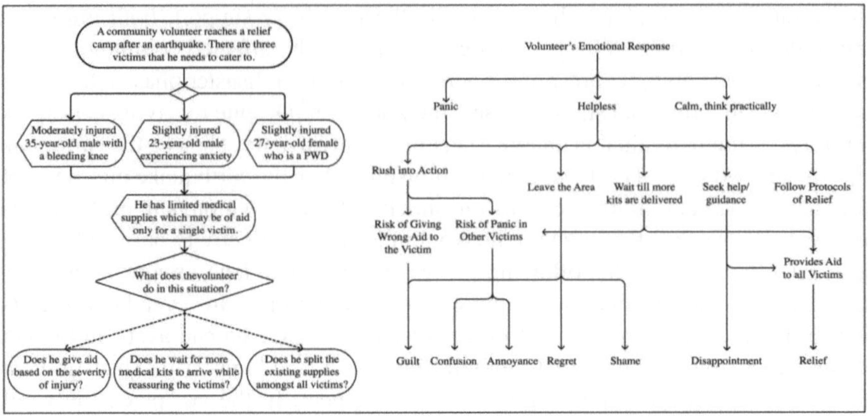

Fig. 3.5 Scenario and emotional decision tree of a community volunteer during an earthquake

knee, the second one is a slightly injured 23-year-old male experiencing anxiety, and the third one is a slightly injured 27-year-old female with a physical disability. The volunteer has limited medical supplies that may be of use for only a single victim. Now, what would a volunteer do in this grave situation? What kind of dilemmas would he face? Would he provide aid based on the limited protocol training he has received? Would he provide aid based on his personal beliefs about age and gender? Would he provide aid based on the ties the victim has to his community? While the scenarios provided an insight into the possible ethical dilemmas the stakeholders would face, emotional decision trees were developed to understand the range of emotional responses they would have in such situations. For example, the community volunteer mentioned in the scenario (Fig. 3.5) may panic or feel helpless, each leading to different consequences. Each action taken in response to the emotion he feels would then also stir up a different set of emotions. These decision trees help map out how emotions influence actions, highlighting potential ethical dilemmas and guiding better training for responders.

When emotions take the lead in high-pressure situations, they could either cloud judgment or guide it toward more humane choices. To understand this further, the emotional responses triggered in such situations were explored. Each situation sparks an initial emotional reaction, shaped by factors like time pressure, cognitive load, emotional intelligence, and experience. Building on the insights from the emotional decision trees, how emotions influence decision-making in disaster contexts was examined. A flowchart (Fig. 3.6) was developed to visualise how this initial emotional response transitions into emotional regulation, a critical stage that determines the kind of decision one makes, whether impulsive or balanced. The nature of emotional regulation is further shaped by context-specific factors such as the severity of the scenario and the urgency of the decision. If emotional regulation is insufficient or bypassed (often due to high stress or time pressure), the individual is more likely to make an impulsive decision (aligned with Fast Thinking processes), which increases

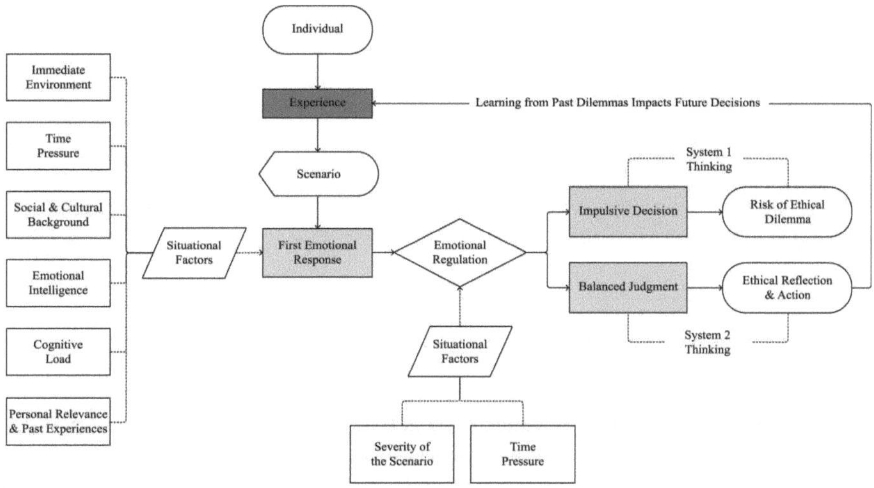

Fig. 3.6 Role of emotions in ethical decision-making: a dual-process model

the risk of ethical dilemmas. On the other hand, effective regulation enables balanced judgment (aligned with Slow Thinking processes), leading to ethical reflection and action [7]. Additionally, learning from past ethical dilemmas feeds back into the individual's experience and influences future decision-making, creating a continuous loop of ethical growth and emotional awareness.

The second set of insights led to examining the effect of stress on performance. Stress plays a crucial role in this decision-making process. It can be described as a body's natural response to challenges. In a situation such as this, where individual effort and group effort are equally important, considering stress only at an individual level would be limiting. For this purpose, a framework (Fig. 3.7) that illustrates how stress impacts decision-making and performance at an individual and, consequently, at a group level, particularly in high-stakes environments like emergency management, was adapted [9]. The primary stressor in this situation was considered to be an event of an ethical dilemma. An ethical dilemma by itself would have other constraints, like time, and a decision-making loop acting on it. The adapted framework integrates individual, team, and organisational dimensions, revealing how stress from ethical dilemmas cascades through these levels to affect overall team performance. It also emphasises how ethical dilemmas exacerbate time pressure, leading to decision delays, increased urgency, and heightened pressure, potentially compromising decision quality. With the help of this diagram, an attempt was made to illustrate the complex interplay between ethical dilemmas and concurrent stressors, which can compound the adverse effects on performance.

To understand the complexity of the system further, all the parts were collated into a gigamap. This led to the identification of patterns, connections, and relationships, both direct and indirect, between different elements. This process helped uncover underlying structures, interdependencies, and potential areas of influence within the

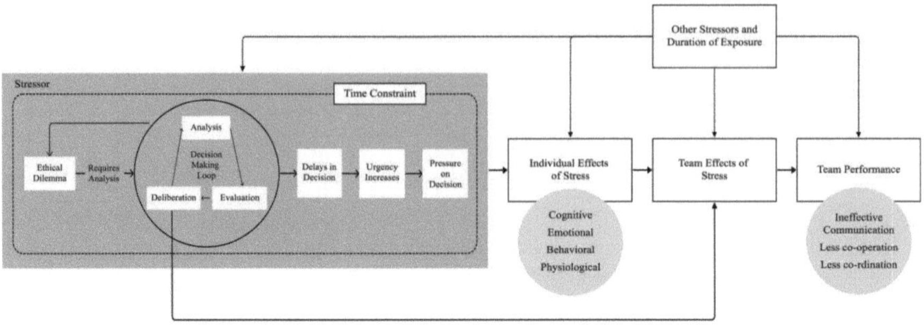

Fig. 3.7 Effects of stress on performance. Adapted from [9]

system, ultimately providing a richer perspective on how it functions and evolves. Through the stakeholder dimensions tool, various characteristics of the stakeholders were analysed, one of which was operational efficiency. This characteristic directly influences their role under stakeholder actions during the relief phase, particularly in terms of resource allocation. However, resource allocation is often affected by limitations that hinder earthquake relief, such as inaccessibility to affected areas and misinformation. These limitations can lead to the emergence of an ethical dilemma, especially when responders are forced to make difficult choices under pressure.and human performanceand performance.

One of the key factors influencing these ethical dilemmas is the presence of organisational and systemic barriers, such as specific protocols or even decision-making structures. These barriers may constrain the ability of individuals within these structures to make flexible, context-sensitive decisions. This chain of events can ultimately result in unethical decision-making, even if unintentionally. For instance, a short-term consequence might be insufficient or misdirected resource allocation, leading to unmet needs on the ground. In the long term, such decisions could result in loss of public trust, damaged credibility, and a diminished reputation, not just for the individual, but for the organisation as a whole. These consequences feed back into the system by affecting future preparedness, cooperation with communities, and overall stakeholder dynamics, thereby extending the web of influence and interdependency. This is just one example of many such inferred relationships that were mapped and explored through the gigamap, revealing the deeply interconnected and cyclical nature of ethics in earthquake relief.

Moving from the granular details to see the larger structure of the system, an influence map was created. This visualisation offered a bird's-eye view of the entire system (Fig. 3.8), allowing one to visualise how different components are interlinked and interdependent. It provided clarity on how the system behaves as a whole, rather than just in parts, revealing overarching patterns, flows of influence, and points of convergence. For instance, a concept like decision-making heuristics doesn't exist in isolation; it influences and is influenced by multiple other elements. It directly impacts the factors influencing ethical decisions, the role of emotions in decision-making, the dual-process thinking model, and even the development of emotional decision

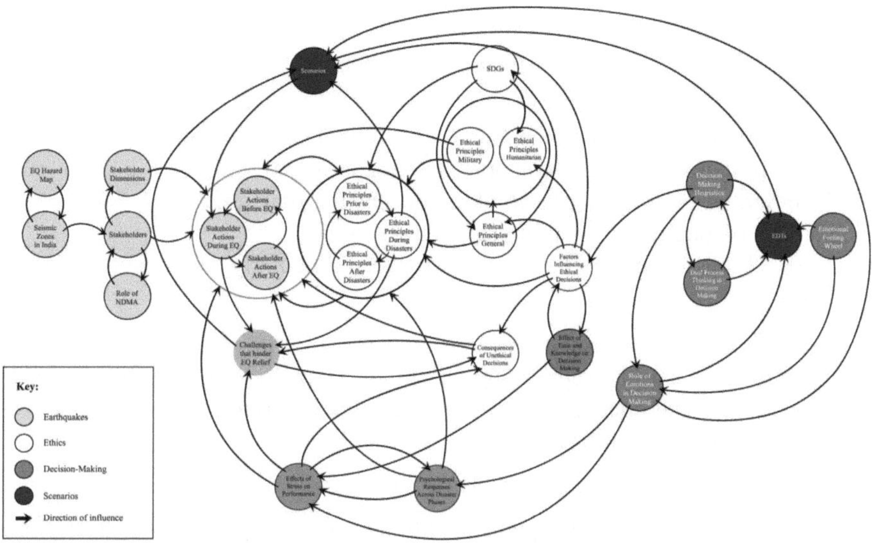

Fig. 3.8 Bird's-eye view of the system

trees. By mapping out these interrelations, the influence map helps visualise how a single component can have cascading effects throughout the system, reinforcing the need for ethical awareness and systemic thinking in disaster response.

3.4 Envisioning the Interventions

The next stage in the process was to envision meaningful interventions, ways in which the system could be strengthened, supported, or ethically improved. To do this, the ZIP Analysis framework was used. Four key elements within the earthquake system were zoomed in on:

- Primary stakeholders.
- Stakeholder actions during an earthquake.
- Limitations that hinder effective relief.
- Psychological responses experienced across the different phases of a disaster.

From this exploration, the Problem Points, the core areas where issues consistently arise, were refined. These included:

- Stakeholder actions during an earthquake.
- Challenges that hinder earthquake relief.
- Conflicts between humanitarian and military frameworks.
- Factors influencing ethical decisions.
- The biases introduced by decision-making heuristics.

Then, the Intervention Points, areas within the system that hold the capacity for ethical growth, systemic improvement, or strategic leverage, were identified. These included:

- Actions by both government and non-government stakeholders.
- The guiding role of ethical principles in disaster response.
- The influence of dual-process thinking in decision-making.
- NDMA's pivotal role in shaping national protocols.
- The psychological phases following a disaster, particularly the Impact and Heroic phases.

Through the rich landscape of problems and possibilities, five key interventions were identified, each one representing a strategic opportunity to enhance ethical outcomes, build resilience, and support more humane disaster response practices. The interventions are as follows:

1. Intervention One—Reducing the Time for Psychological First Aid.

In most cases, psychological support is only introduced in the post-disaster phase, often days after trauma has occurred. This delay can significantly affect long-term mental health. Here, the proposed shift was delivering Psychological First Aid within hours, during the immediate response phase. This intervention directly supports victims, as well as first responders, volunteers, and medical personnel, ensuring emotional care is prioritised from the very beginning.

2. Intervention Two—Ensuring Ethical Integrity in Relief Efforts

Speed in crisis response is vital, but often, ethical considerations are overlooked in the rush. This intervention focused on embedding ethical decision-making awareness into disaster preparedness. By equipping volunteers, responders, and probable victims with the tools to navigate ethical challenges, we aimed to foster decisions that are not only fast but fair, inclusive, and humane.

3. Intervention Three—Enhancing Coordination in Resource Distribution

Earthquake relief efforts are often hindered by inequitable and inefficient resource allocation—some areas face shortages, others surpluses. To tackle this, a dynamic, real-time tracking system for distributing critical aid based on verified needs and accessibility was proposed. This ensured that aid would be delivered transparently and equitably, reducing hoarding, bias, and administrative inefficiencies. Stakeholders here include volunteers, NGOs, authorities, and the affected communities themselves.

4. Intervention Four—Real-Time Updates on Missing Persons

Fragmented communication systems cause delays, confusion, and distress for both families and relief teams. A real-time, publicly accessible platform that streamlines the process of reporting and locating missing persons was envisioned. This transparency would help reduce duplicate search efforts, support quicker reunification, and minimise ethical dilemmas related to resource prioritisation. Key users would include victims, volunteers, coordinators, and local authorities.

5. Intervention Five—Bespoke Household Preparedness Modules

Preparedness today is largely generic and top-down, and often ignores the lived realities of individuals in different geographic and socio-economic environments. This intervention called for customisable, context-specific modules designed for households, empowering families to take ownership of their safety with clear, adaptable strategies. This bottom-up approach supports households, local communities, and strengthens systemic resilience at the grassroots level.

Intervention points with the corresponding leverage point, anticipated impact, and potential risks, including revenge effects that could arise from unintended system responses, were then mapped out. Starting with the first intervention, the leverage point here was the timing of the intervention; a simple change in process can significantly improve trauma recovery outcomes. This centres emotional care in the response timeline. However, if poorly implemented, this could lead to misuse by untrained personnel or erode trust if victims feel neglected or mishandled. The second intervention targets a mindset shift in decision-making under pressure. This promotes fairness and inclusion, embedding ethics into practice. But over-regulation could backfire, slowing response times or creating conflicts among field teams.

The third intervention leveraged transparency and adaptability to reduce inefficiencies. While it could drastically improve trust and aid accuracy, revenge effects might include manipulation, hacking, or even political misuse of the system. The fourth intervention would strengthen information feedback loops, reducing ethical dilemmas and emotional toll, and adding coordination and transparency. While it promises informed rescue and emotional reassurance, it also introduces risks of data breaches, privacy violations, and the spread of misinformation. Lastly, Intervention 5 supported community decision-making and localised resilience. The shift here would have been from top-down mandates to bottom-up agency. The risk lay in over-reliance on households, potentially neglecting necessary systemic infrastructure.

3.5 Formulating the Interventions

Each of these interventions held high potential but also carried complex system interdependencies. Among the five proposed interventions, 'Intervention 2—Integrating Ethical Awareness into the preparedness phase' was selected to be further developed. It was selected for its potential to create systemic change at the grassroots level, directly influencing the mindset and behaviour of those involved in the frontline of the disaster, including volunteers, first responders, relief coordinators, and probable victims. Rather than being reactive, this was a precautionary intervention. It would prepare stakeholders in advance, equipping them to handle high-pressure, emotionally charged situations with clarity and fairness. In doing so, it would transform ethics from being an afterthought into an embedded, actionable practice. Another reason this intervention stood out was its scalability. It would reach a broader audience through structured awareness, cascading its impact across regions and responder networks. And while it may seem like a small change in content, it holds the power to drive a much larger systemic impact, improving fairness, inclusivity, and trust in relief efforts.

To bring this intervention to life, a public awareness campaign (Fig. 3.9) focused on promoting ethical decision-making in earthquake relief was proposed. This campaign aimed to equip both relief workers and communities with the understanding and tools needed to make fair, inclusive, and responsible choices in moments of crisis. The campaign would embed ethics in disaster preparedness, ensuring that aid reaches those who need it most, without bias or oversight. It would also highlight the rights and dignity of survivors, reinforcing a culture of integrity, collaboration, and accountability. By encouraging cooperation between responders and communities, the campaign would foster shared responsibility in ethical decision-making. Ultimately, this solution would lead to more transparent, consistent, and compassionate relief operations, improving both immediate response and long-term recovery outcomes.

The campaign was designed for two primary groups within high-impact earthquake zones—probable victims and disaster responders. On one side were the earthquake-affected communities, families, and individuals who might find themselves in crisis, often with limited access to ethical guidance or accurate information. On the other hand, those involved in relief efforts, such as volunteers, NGO workers, first responders, and medical personnel, who are often forced to make quick, high-stakes decisions under pressure, were addressed.

The campaign would take distinct forms in urban and rural areas, based on each setting's social structure, communication channels, and behavioural patterns. To

Fig. 3.9 Poster samples from the proposed awareness campaign

understand the needs and pain points of the involved individuals, three representative personas were developed. They are described as follows-

- Ananya, a young urban professional, lives alone in a high-rise in Delhi. She lacks awareness about ethical disaster relief and may unknowingly spread misinformation in a crisis.
- Rohan, an experienced NGO volunteer, regularly coordinates urban relief but often faces ethical dilemmas, like deciding how to distribute limited aid fairly while navigating media pressure.
- Suman, a nurse in rural Bihar, works tirelessly in under-resourced conditions. She frequently must make life-or-death decisions without clear ethical guidance, especially when dealing with vulnerable, less-informed populations.

These personas reflected the diversity of ethical challenges faced across different contexts and highlight why tailored, ethics-driven awareness is crucial.

In urban areas, the campaign will unfold in three phases—The Launch phase would ensure high visibility through posters and flash mob performances that dramatise ethical dilemmas in crisis scenarios. The Engage phase would involve inclusive community sessions where survivors, first responders, and policymakers share real experiences and reflections. While the Reinforcement phase would use flyers, interactive competitions, and info modules to keep the conversation going, it will cement the message through repetition and creativity.

In rural areas, the approach respects local culture and community rhythms—The Launch phase would start with awareness sessions near village gathering points like temples, anganwadis, and markets, supported by flyer distribution. The Engage phase leverages local traditions like nukkad nataks and folk songs to communicate ethical issues in an accessible way. Community workshops would help surface real stories and shared wisdom. And the Reinforcement phase would embed the message in everyday life: with murals, zines in relief kits, and musical storytelling that would live on in memory and word-of-mouth.

3.6 Enabling the Interventions

This phase looks beyond implementation, focusing on how the solution gradually becomes part of the system, how it would be accepted, sustained, and ultimately transform the way people act, decide, and prepare. To accomplish this, a Roadmap for Transition to visualise how the campaign would mature from an idea into a lasting impact was used. The roadmap follows a transition pathway, moving from the current state to a preferred future. At the starting point, there is minimal awareness of ethical decision-making in disaster response, and decision-making often defaults to urgency rather than fairness. With the public awareness campaign as the enabler, interventions in urban and rural contexts would be initiated. These would roll out in three key stages: Launch, Engage, and Reinforce, each adapted to suit different community behaviours, media channels, and trust systems. As the campaign would unfold, the phase of community adoption, where individuals begin to internalise ethics as part of preparedness, not just crisis management, is there. This would happen through repeated exposure, interactive tools, and local dialogue. From here, we move toward systemic embedding, where ethical principles start influencing how volunteers are trained, how relief is coordinated, and how decisions are made under pressure.

Ultimately, the goal was a long-term culture shift where ethical decision-making becomes intuitive, collaborative, and embedded in the DNA of disaster response systems. The potential long-term impact of this intervention lies not only in improving outcomes during disasters but in changing how people think and act long before a disaster occurs. By embedding ethics into the mindset of both responders and communities, the groundwork is laid for a more just and equitable disaster response culture. This campaign could influence training programs, policy design, NGO protocols, and even how survivors advocate for themselves. Over time, these micro-shifts build toward a more resilient and empathetic society—one that is prepared, informed, and grounded in shared values when disaster strikes. To ensure the campaign achieves this vision, a cyclical impact evaluation model can be adopted—Act, Observe, and Decide. During the Act phase, activities are rolled out to build engagement. During Observe, reach would be tracked, participation data would be gathered, and the community would be heard through surveys, feedback sessions, and field observations. Then, in Decide, the messaging would be refined, formats would be adapted, and distribution would be improved, making the campaign more responsive, context-specific, and effective over time. This cycle would allow the campaign to evolve as the landscape changes, ensuring that ethical awareness wouldn't just reach people but would resonate and endure.

3.7 Conclusion

Wicked problems, like those in disaster response, don't come with clear boundaries or easy solutions. They demand more than quick fixes; they require ethical imagination, systemic thinking, and an openness to learn from uncertainty. This systems analysis project explored the problems that are currently present in earthquake response in India. The future direction of the project would be to analyse alternative ways of bringing a mindset shift at the grassroots level and expanding the suggested solution into an adaptable training program for different stakeholders.

References

1. Ministry of Earth Sciences (2025) Making India more disaster-resilient: Government's proactive measures for earthquake safety
2. Yates JF, De Oliveira S (2016) Culture and decision making. Organ Behav Hum Decis Process 136:106–118. https://doi.org/10.1016/j.obhdp.2016.05.003

3. Prieur M, European and Mediterranean Major Hazards Agreement (EUR-OPA) (2009) Ethical principles on disaster risk reduction and people's resilience
4. Innovative methods to save people from disasters. https://www.pib.gov.in/PressReleseDetailm.aspx?PRID=2110352®=3&lang=1
5. National Disaster Management Authority, Government of India (2016) National disaster management guidelines: hospital safety
6. National Disaster Management Authority (2008) National guidelines for management of cyclones. Government of India, New Delhi. https://www.ndma.gov.in/ndma-guidelines. Accessed 19 Mar 2026
7. Dale S (2015) Heuristics and biases. Bus Inf Rev 32:93–99. https://doi.org/10.1177/0266382115592536
8. Math SB, Nirmala MC, Moirangthem S, Kumar NC (2015) Disaster management: mental health perspective. Indian J Psychol Med 37:261–271. https://doi.org/10.4103/0253-7176.162915
9. Butler PC, Flin R, Bearman C, Hayes P, Penney G, McLennan J (2024) Emergency management decision-making in a changing world: 3 key challenges. Aust J Emerg Manag 39(4):23–32. https://doi.org/10.47389/39.4.23

Chapter 4
Indian Millets and Their Ecology: Interventions for a Healthier Future

Parnavi Dinkar, Rahul Raj Singh, and Tina Soni

Abstract Millets are climate-conscious crops that require less water, making them an integral part of future food systems. Despite the established drawbacks of wheat-rich diets, they dominate the market while millets remain underrepresented. This can be attributed to low discoverability, lack of consumer awareness and limited product innovation. The project objective was to redirect consumers' focus towards indigenous millet-rich diets, while considering the geography, climate, demographics, and culture. The project began with examining the cooking process and the type of breads traditionally prepared in different regions of India. This revealed the underlying relationship between climate, agricultural processes, and local diets, with respect to millets. The interdependencies between industries in the system were identified. The emergent factors that directly and indirectly contribute to the decline in cultivation, distribution and consumption were organised using stakeholder mapping, stakeholder dimensions and classification tools. The intervention points were identified by zooming in on integration, literacy, supply chain, products and digestion of millets. This led to the identification of awareness as the leverage point and the development of millet experiential stores as a solution. The proposed solution focuses on incorporating millets in mainstream diets by shifting the narrative of millets as the grain of the future. An increase in awareness would bring a change in consumer behaviour, thus paving the road for the integration of millets in regular use. The proposed solution also aligns with the Sustainable Development Goals of Good Health & Well-being (Goal 03), Responsible Consumption & Production (Goal 12) and Climate Action (Goal 13).

Keywords Millets · Consumption · Awareness

P. Dinkar (✉) · R. R. Singh · T. Soni
Indian Institute of Technology Kanpur, Kanpur, India
e-mail: parnavid24@iitk.ac.in

V. Kant (ed.), *Human Centred Design for Systems*, Springer Series in Design and Innovation 68, https://doi.org/10.1007/978-3-032-30376-9_4

4.1 Introduction

Millets offer a more sustainable, nutritious, and climate-resilient alternative to other resource-intensive grains. They require less water, are pest-resistant, making them a crucial component of the food ecosystem. Despite being a crucial and beneficial component of the food ecosystem, the usage and perception of millets are limited. One of the factors influencing the low visibility of millets in the market is the limitations faced during production. Due to the economic and financial pressures, farmers struggle to support themselves solely on millet production and are compelled to grow cash crops. Another factor that influences the public's accessibility to millets is the disparity in price. Rice and wheat are distributed for free or at nominal prices by the public distribution system, whereas millets are sold at subsidised but higher rates in comparison [1]. Millets have been used to prepare bread and in the cooking process since as early as the Harappan civilisation [2]. Over a period of time, changing cooking methods and commercialisation have brought about a change in the manner in which millets are consumed.

4.2 Framing the System

Cooking is a fundamental human activity that is rich in regional variations, rituals, and resources. Cooking habits tend to evolve, are sustained, or tend to fade over time. The Cooking and consequently the food that is prepared are influenced by culture, the history of the region, the available technology, and the policies in effect. The evolution of cooking in India is also largely influenced by accessibility and adaptation. The nature of cooking practices has changed across eras, from ancient and medieval techniques to colonial influences and post-industrial habits. These habits shape the understanding of people—people are what they eat. This chapter explores the different facets of cooking, specifically focusing on millets as a source of nutrition in the Indian context.

This study started with a broad lens, initially examining 'Breads'—one of the staple foods considered across various regions of India. This food category is quite common, yet very diverse. Breads, being more than recipes, also reflect the overall climate, crop cycles, tools, and people's digestion patterns.

4.3 Comprehending the System

The cycle of manufacturing to consumption of bread involves multiple stakeholders, right from individuals and groups who cultivate the crops to those who sell and consume the end product. On the basis of their role and involvement in the Indian Bread-making system, the stakeholders were divided into primary, secondary, and tertiary stakeholders (Fig. 4.1). Primary stakeholders were considered to be those directly involved in the immediate production or consumption of bread. The secondary stakeholders were considered as those who did not participate in the immediate production or consumption process but played a vital role in linking the primary stakeholders with the system. For example, culinary schools that teach bread preparation, dieticians who recommend bread as part of a balanced diet, or storage facilities that preserve grains were categorised as secondary stakeholders. While stakeholders such as pesticide manufacturers, who play a crucial role in ensuring the successful cultivation of grains but have no direct involvement in the production or consumption process, were defined as the tertiary stakeholders. Here, the pesticide manufacturers are not directly linked to bread production; however, their role in supporting farmers who grow the necessary crops used in bread is crucial, as the absence of such agricultural inputs would cause grain production to suffer, consequently affecting the bread supply chain and the system at large. The relationship among the stakeholders was analysed, and clusters were formed based on the commonality of their roles, such as raw material providers, tool makers, distributors, and service providers. This categorisation led to the identification of variables that influence the system.

In order to understand the degree of impact the variables have on the system, they were mapped based on their defining characteristics. One such example of the variable would be 'Manufacturing of Farming Tools'. The defining characteristic considered here was 'Size of Farming Tools Manufactured', which ranged from a small screw to a combine harvester, one of the largest agricultural machines. Some variables, however, could also have multiple defining characteristics. For instance, if a 'Cook' is considered to be a variable, then its defining characteristic could be both 'Type of Training', which could range from 'Self Trained' to 'Formally Trained'.

Along with understanding the level of impact the variables would have on the system, exploring the variety of bread produced in the Indian context was of equal importance. This exploratory exercise revealed distinct types of breads cooked across the country, each unique in its own way (Table 4.1) [3]. Some bread types shared similarities in their preparation style or the final product but often had different names depending on the region they were from, making them uniquely tied to their local cultures. These types were categorised based on their state of origin as well as based on their form and distinctive textures, such as flatbreads, steamed cakes, pancakes, crepes, and others. This analysis highlighted the diversity of Indian breads, as well as the unique identities formed due to the regional variations within the culinary landscape.

Smoky roomali rotis, earthy makki di rotis, and leavened kulchas cooked in tandoors were primarily found in the northern parts of India. Towards the west, theplas, khakras and rotlas, breads that are travel-friendly complementing the region's dry climate, were found. Litti, malpua, and pitha (stuffed or steamed breads tied to festive customs) were predominantly found in the central and eastern parts of India. While in the south, akki roti, adai, appam, idiyappam, and puttu (fermented, steamed and pan-cooked breads) were found. In addition to the categorisation as per regions, another set of classification was done based on the factors of form or the bread,

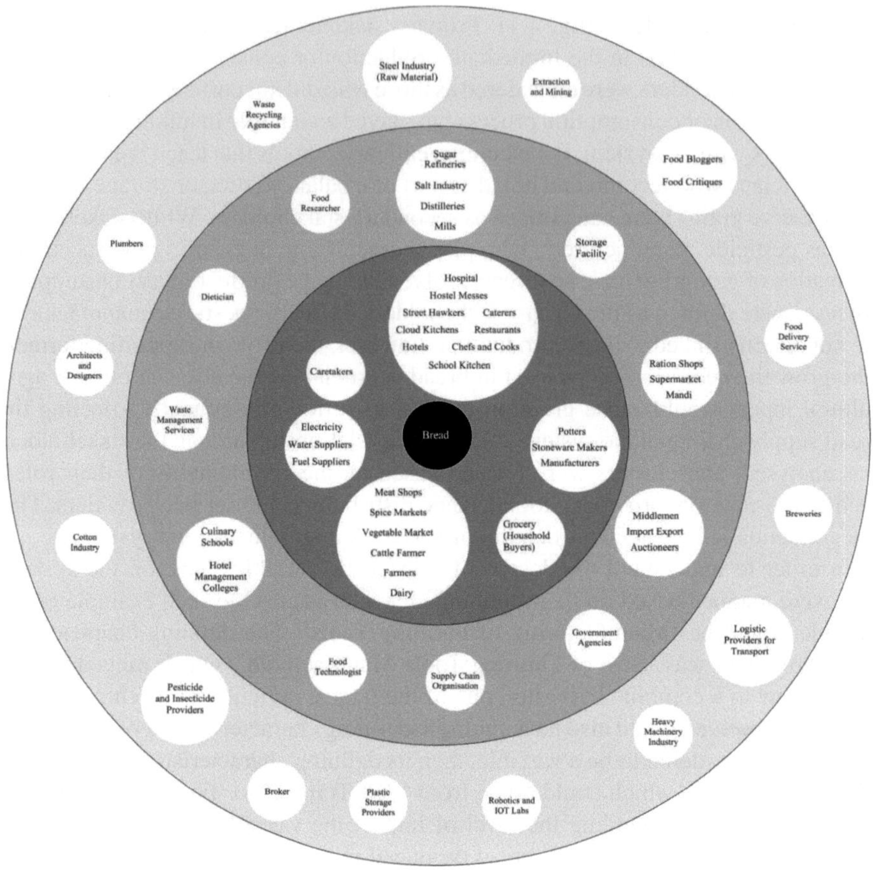

Fig. 4.1 Stakeholder mapping

texture of bread, type of bread, core grain used, and the historical period in which the bread originated [3].

To understand the diversity in breads, further exploration focusing on the process of making the bread was carried out. Here, three core categories emerged: Flatbreads, Steamed Cakes, and Pancakes. Flatbreads are those made from rolled dough grains, such as wheat, millet, or maize, mixed with water and sometimes dairy, flattened by hand, and cooked dry on a griddle or tandoor. For example, phulka, luchi, makke di roti, koki, or sheermal. These are the breads that are most often seen in daily Indian meals, which are fast to prepare, filling, and adaptable. Pancake-style breads are prepared with a batter. These breads are thin, soft, sometimes fermented, and cooked on a flat pan, e.g. dosa, uttapam, appe, or malpua. Steamed cakes such as idli, puttu, idiyappam and sanna, which are light, fluffy and easy on the stomach, are cooked by pouring batter into moulds and steaming them till they rise. These breads are often paired with curries or gravies, balancing out their mild taste with

Table 4.1 Types of bread [3]

State/Region	Type of flatbread(s)
Jammu & Kashmir	Girda, Taftan, Sheermal, Bakarkhani
Himachal Pradesh	Siddu
Punjab	Bhatura, Kulcha, Makki di Roti, Rumali Roti
Uttar Pradesh	Khamiri Roti, Lachcha Paratha
Bihar	Litti
Assam	Pitha
Manipur	Tan Ngang
Mizoram	Pukhlein
West Bengal	Luchi, Gol Gappe
Odisha	Malpua
Andhra Pradesh	Adai
Tamil Nadu	Idiyappam, Dosa, Appe
Kerala	Parotta, Appam
Karnataka	Akki Roti
Goa	Pao
Maharashtra	Bhakri, Puranpoli, Sarva Pindi
Gujarat	Koki, Thepla, Khakhra
Rajasthan	Papad, Baati, Missi Roti, Rotla
Madhya Pradesh	Phulka

rich accompaniments. These three types of flatbreads, steamed cakes, and pancakes were kept flexible to provide a clearer view of the system. This helped clarify why certain breads suit individuals better from different places. One of the factors that plays a critical role in this system is climate. In hot, dry zones where rainfall is scarce, tough grains such as bajra and jowar, and breads like rotla, khakhra, or baati are found. These store well and do not require large amounts of water to be cooked. In high-rainfall zones such as Bengal or Kerala, rice-based breads are found, which are either steamed or fermented. They are found to be gentle on the gut and ideal for humid conditions. In a temperate belt like this, mixed grain breads are made with jowar, wheat, chana dal, or rice, depending on the rainfall [4]. The next facet of the system that was explored was the tools and equipment used in bread-making. From traditional tavas, clay ovens, and steaming baskets, to modern-day non-stick pans and ovens, the diversity of bread is also influenced by the variety of cooking equipment utilised. Each tool is optimised for the grain and method. Vice versa, the cooking method also evolves, adapting to the tool. A bhakri pan handles millet better than a standard tawa, whereas dosa pans require even spread heat and minimal oil.

The research showed that most Indian households use the tava (a flat, usually iron skillet) to make daily rotis. For millets like bajra or jowar, thicker, heavier pans are preferred that retain heat for longer periods. This is because the millet flour is harder to roll, and careful attention is required for the millet to be cooked evenly. Home kitchens are all about adaptability. A flatbread might be cooked over an open flame, flipped by hand, or puffed directly on the stove. A pancake-style bread might be made in a non-stick pan. These dynamics change in a commercial kitchen. In hotels, canteens, or even community kitchens, standardisation of food is required on a larger scale. Large gas griddles, tandoor ovens, rotary baking plates, and automated roti makers are used in such kitchens. Processes such as steaming of idlis are carried out in industrial batch steamers that can contain large batches of idlis at a time. This difference between home and commercial cooking equipment is part of the reason why certain breads have become more mainstream compared to others. Wheat chapatis are easier to scale, store, and mechanise. Bajra rotis or fermented millet dosas, on the other hand, are more delicate and therefore are less likely to be adopted in institutional kitchens. The selection of equipment is not limited to the cooking process, but expands to the required capital, added effort and necessary skillset. Among the core grains used to make the bread, "Millets" emerged as an important area of exploration as they offer a considerable amount of health benefits, yet their usage and awareness were found to be limited. Millets tend to grow where other crops often struggle to survive. Table 4.2 illustrates the quantity of millet being produced in the corresponding state [5]. Millets adapt to thrive in regions with poor soil, irregular rainfall, and harsh climates.

Along with understanding where millets are cultivated, understanding what happens to the food when it is consumed is also important. The human digestive system was studied to understand how different ingredients used in breads are broken down into glucose and fibre, while other nutrients get absorbed by the body and are eventually excreted [6]. Millets and fermented options can be considered as an alternative to be used for making healthier breads.

In the early years of development, children require nutrient-dense foods that are gentle on the gut. Millets check both boxes. The high fibre, iron, and protein content support growing bodies, while being gluten-free reduces the risk of early

Table 4.2 Largest millets producing states in India 2025 [5]

State	Share of total millet production (%)
Rajasthan	27
Karnataka	18
Maharashtra	14
Uttar Pradesh	12
Gujarat	7
Madhya Pradesh	6
Tamil Nadu	5
Andhra Pradesh	4
Uttarakhand	2
Haryana	1

gut inflammation or allergies. In adulthood, the body's metabolic and digestive efficiency is at its prime. But this is also the phase where poor eating habits solidify. Fast food, processed carbs, and nutrient-poor meals lead to bloating and similar issues. Including millets in adult diets offers a way to balance this out. Their slow-releasing carbohydrates help maintain energy levels, reduce spikes, and aid in long-term gut health. In older adults, the body starts slowing down. Digestive enzymes weaken, and gluten sensitivity becomes more common. This is where millets are beneficial. They are lighter, easier to break down, and rich in fibre, which helps in regulating digestion and supporting nutrient absorption. They're a timeless food that suits the body at all ages. That makes them not only nutritious but necessary.

Both millet and rice are high in carbohydrates. However, that is where the similarities end. Millets provide significantly more protein. Nearly 13 g of protein per 100 g of food, which makes it far more filling and useful for energy and cell repair [7]. It also offers more fat, which helps with the absorption of fat-soluble vitamins. Millet also has almost three times the fibre of rice, which keeps the gut moving and the microbiome healthy. In addition to that, calcium, iron, and a range of micronutrients are also available in millets. Rice, in comparison, is lighter and easier to digest in some cases, but nutritionally, it is thin with a lower content of fibre, less protein, and minimal micronutrients. In the case of wheat and maida-based breads, the challenges are medical. Over time, an increasing number of people have reported discomfort after eating gluten-rich wheat breads. What starts as simple bloating or fatigue can sometimes escalate into more serious conditions like non-celiac gluten sensitivity or even full-blown wheat allergy. The symptoms vary headaches, rashes, digestive pain, and fatigue. For others, especially those with celiac disease, gluten becomes a true threat. It triggers inflammation, damages the gut lining, and leads to nutrient malabsorption that can further develop into stomach aches, weight loss, osteoporosis, long-term fatigue, and sometimes, irreversible damage [7].

The process of digestion does not work in isolation from the external factors. It also connects to geography. What grows in a region is what people have traditionally eaten. And that, over time, shaped what their bodies are used to digesting. But today, due to migration and urbanisation, people tend to eat in ways that don't align with their climate (Fig. 4.2). When people move from one location to another, their food habits don't change to a great extent. As a consequence, due to the difference in climate, the ingrained food habits might be unsuitable in the new location. The first generation that migrates tends to continue to consume the foods they are familiar with, while the second generation starts eating whatever is easy to access [8].

The use of millets in the preparation of Indian breads has been limited. For a country that once depended on it, both agriculturally and culturally, millet's decline was influenced by a Western-driven focus on refined grains. Millets are nutritious, climate resilient, require less water and are good for the environment. Most importantly, they are a healthier choice and can be substituted for refined grains [9]. However, despite the evident benefits, there is a lack of awareness among the general population about millets. Millets consist of an entire family of jowar, bajra, ragi, kodo, and more, each with its own taste, texture, and nutritional profile (Table 4.3). Ragi is great for calcium and bone strength. Bajra keeps the body warm and is high

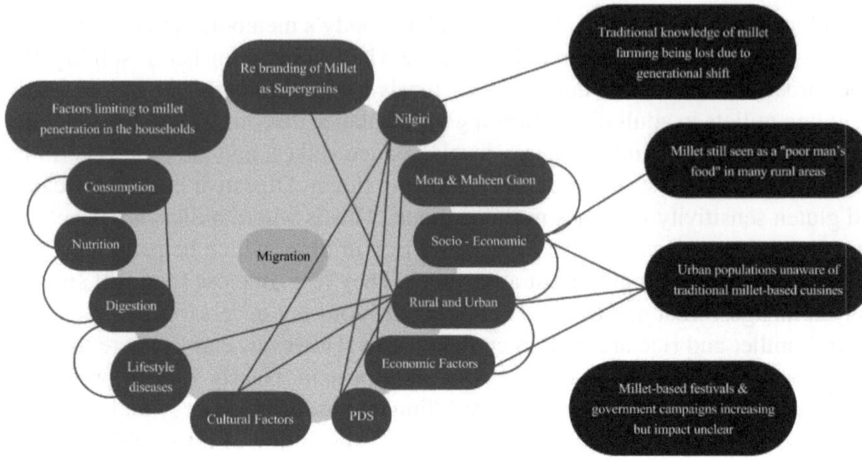

Fig. 4.2 Migration of millets [8]

in iron. Kodo supports blood pressure and purifies the blood. Little millet supports thyroid health. Each millet is a different nutritional packet. But in the mainstream food narrative, they've all been reduced to the 'Poor Man's Grain'.

In rural India, millets are consumed out of necessity. They're locally grown, available, and affordable. However, this perception is starting to shift, because rice and wheat have better subsidies. Simultaneously, the younger generations are opting for packaged foods. In urban areas, millets are seen more as a health experiment or a trend, but not something to rely on. This is due to the emotional and cultural familiarity with wheat and rice over millets. The current food landscape thus shows the rural populations stepping away from millet out of aspiration, while the urban populations approach millet out of curiosity. Neither of the groups is fully rooted in it.

The decline of millet in India began with the revenge effects of the policies of the 'Green Revolution'. While it was necessary at the time, it focused on wheat and rice to combat famine by introducing high-yielding varieties, investing in irrigation, and stabilising food security, which made rice and wheat easier to grow. But in doing so, it pushed millets out of the frame. Then came the Public Distribution System,

Table 4.3 Different types of millets

Millet	Types of millets
Major millets	Sorghum (Jowar), Pearl Millet (Bajra), Finger Millet (Ragi)
Minor millets	Foxtail Millet (Kakum), Kodo Millet (Kodon), Barnyard Millet (Kodon), Little Millet (Kutki), Proso Millet (Chenna)
Pseudo millets	Amarnath (Rajgira), Buckwheat (Kutku)

where rice and wheat were also prioritised. They were easier to store, easier to cook, and could be given away for free. But millets were harder to process, took longer to cook, and were excluded from most state-level PDS programs. Even today, only a few millets, like ragi, jowar, and bajra, have been introduced in government schemes. As a consequence, millet cultivation and awareness declined. Census data show that while the fall isn't catastrophic, the numbers are lower than they were before the famine era. And that, in itself, is alarming, as it is not just a decline in numbers, but a decline in the perceived value [10].

To understand why millet isn't as commonly consumed anymore, revisiting its history is important. Traditionally, millets were dietary staples, sustainably grown and locally eaten. But due to inefficient agricultural planning, food shortages became a national concern. That is when the Green Revolution was introduced [11]. With its promise of food security, it introduced high-yielding wheat and rice crops, heavily subsidised by the Public Distribution System. This shifted the entire narrative of what a staple grain should be. Millet was marginalised as it did not fit the new model of industrial farming [11]. Over time, the preference shifted from what was climate-resilient to what was easier to grow, cook, and store. This shift in policy and preference slowly created a ripple effect that changed the urban diets and eventually led to their absence from household foods.

Earlier, India's agriculture was scattered, local, and vulnerable. Low-yield farming, climate unpredictability, and unplanned crop selection meant food shortages were common, and imports became the safety net. When the Green Revolution was introduced, it brought in high-yielding seeds, fertilisers, irrigation, and machines, boosting the production of wheat and rice dramatically. The state of food availability in India shifted from food-deficient to food-surplus. By the 1980s, India was exporting grains. This progress, however, led to an imbalance. Millet grains that required less water, fewer inputs, and gave more nutrition were not used as an alternative. Overuse of chemicals began to degrade the soil, groundwater levels dropped, and regional inequality increased. Despite India becoming self-sufficient, it also became dependent on a narrow band of crops. Today, better food systems are required that also cater to problems related to climate volatility, water scarcity, and rising lifestyle diseases. Millets may be considered a potential solution for these issues.

One of the often overlooked reasons why millet is unpopular in kitchens today is its short shelf life. Unlike wheat or rice flour, which can stay fresh for up to six months, flour made from jowar, bajra, and even ragi tends to go stale much faster, usually in about three to four months. This is due to their higher fat content and lack of processing; they go rancid and taste bitter. What this means is that millet flour needs to be used quickly, stored carefully, and bought more frequently. For urban consumers who are used to buying in bulk or depending on long-lasting pantry staples, this becomes a hassle. And in retail terms, it makes millet less viable for large-scale commercial packaging and distribution. So, even though millet is nutritious, the shorter shelf life has contributed to its declining visibility in the market [12].

Millets often have a hard outer husk that takes effort to remove, especially in varieties like foxtail and barnyard millet. This contributes to the added effort of cooking millet. Grains like ragi need to be softened before use by the method of soaking and boiling, which takes time. They also contain anti-nutrients like phytates and tannins, which reduce mineral absorption if not prepared correctly. The dough, too, is tricky. Due to a lack of gluten, millets don't bind well, making it harder to roll or knead. Then there's the taste, earthy, nutty, sometimes even bitter, especially in kodo millet. These aren't easy flavours to adapt to if you've grown up eating refined rice or wheat. All these factors create friction in everyday cooking, adding to the perception that millet is 'inconvenient' or 'too traditional' for modern lifestyles.

The Nilgiri Revival Project [13] is a good example that accurately explains the current scenario. In the colonial era, the hilly regions of Nilgiri shifted to cash crops like tea and coffee, and millets disappeared. With the influence of PDS, cheap wheat and rice took over local diets. However, the consequences of that shift were dire. Malnutrition, anaemia, and poor gut health became prevalent among the tribal communities. Recognising this, state governments and grassroots organisations started to reintroduce millets under the label of 'nutri-cereals.' In Karnataka, they're known as Siridhanya (noble grains). Through school mid-day meals, awareness drives, and farmer incentives, these projects are not just reviving millet cultivation but also restoring its place in the local food culture [13].

Despite all this, one stereotype that limits millet from becoming mainstream is its image as the "Grain of the Poor." This pyramid, shown in (Fig. 4.3) represents the social perception of millets. On top are rice and wheat, seen as premium grains, dominant in wealthier and urban households. In the middle are hybrid flours or processed blends that serve as compromises, fitting into fast-paced, commercial lifestyles. And at the bottom are millets—coarse, unrefined, and traditionally associated with poverty. This visual hierarchy is about class, convenience, and cultural value. Today, however, millets are gaining popularity among health-conscious urban elites as 'supergrains.' What used to be neglected as poor man's food is now being rebranded as the smart man's choice.

Super grains are nutrient-dense, climate-resilient crops that offer exceptional health and agricultural benefits. Millets, often overlooked, truly embody this definition and are essential for a sustainable, healthy future. Nutritionally, they are powerhouses, high in fibre, iron, calcium, and protein, with a low glycemic index. Unlike many modern grains, millets are naturally gluten-free and support digestive health, diabetes management, and cardiovascular well-being. As they are adaptive, accessible, and have immense health benefits, globally, they are being rebranded as supergrains.

When compared to mainstream grains like wheat and rice, millets emerge as the more sustainable choice, both nutritionally and ecologically. Mainstream grains often undergo refining, stripping them of essential nutrients. Millets, on the other hand, retain their natural fibre, aiding digestion and promoting satiety. They're rich in micronutrients, like zinc, magnesium, and B-complex vitamins, which are essential for maintaining metabolic health. Millets score low on the glycemic index, preventing sugar spikes and helping manage diabetes, which is a major concern in India today [14]. Rice cultivation consumes nearly 4000 litres of water per kg, while millets need only a fraction of that. In times of climate stress and water scarcity, millets prove to

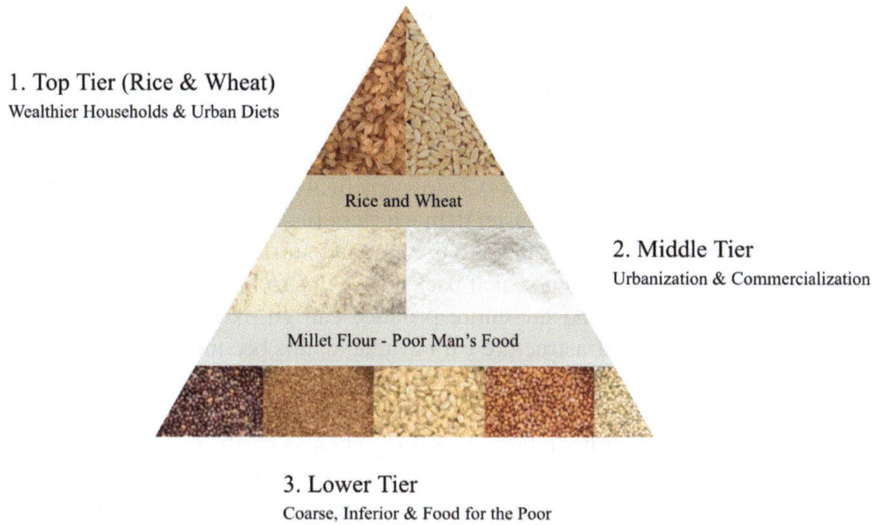

Fig. 4.3 Social perception of millets

be an ecological necessity. Unlike monoculture wheat or rice fields, millet farming encourages biodiversity, improves soil health, and reduces dependency on chemical inputs. From plate to planet, millets outperform mainstream grains, thus making a strong case for future food security.

Apart from being regionally resilient, they're seasonally flexible too. Most millets are grown during the kharif season, from June to September, but some, like ragi and foxtail millet, can even be sown in the rabi season. States like Uttar Pradesh, Rajasthan, Madhya Pradesh, Maharashtra, and Karnataka are leading producers of millets, with high yields and growing government support [15]. Knowing when and where millets are grown helps to better understand how they can fit into larger food systems as a staple food.

The UN declared 2023 the International Year of Millets. Campaigns, workshops, and millet festivals were initiated to increase awareness. Some states are rebranding millets while urban farms have begun incorporating them into their cultivation practices. This has slowly started to shift the perception of millet being a "Poor Man's Grain" to becoming a 'Supergrain'. The key now is to keep this momentum going, to increase millet's visibility and bring it back as a necessity [16].

4.4 Envisioning the Interventions

The important question here persists: if millets are so good nutritionally, environmentally, and economically, then why aren't more people eating them? This is where the problem became more complex. It is a layered mesh of perception, policy, practicality, and price. To start with, one of the most practical issues is "Shelf Life". Millet, in its processed flour form, spoils faster than wheat or rice [12]. After milling, it starts reacting with oxygen, degrading flavour, texture, and nutritional value. If not stored in airtight containers or vacuum-sealed packs, it begins to lose quality rapidly. Add moisture to the equation, and the entire produce can go to waste. In rural areas, where advanced storage infrastructure isn't available, this becomes a major setback. An added layer of complexity is brought in during the cooking process. Unlike wheat, millets don't contain gluten. While this is beneficial for the digestive system, binding millet dough becomes a difficult task. This dough cracks easily, doesn't stretch, and once cooked, it tends to dry out quickly. Unless it is softened with ghee or paired with stews or curries, it can feel coarse or bland to people used to soft wheat chapatis.

Taste is another factor, despite being subjective, that plays an important role in influencing an individual's decision to consume it regularly. For example, when choosing between a soft, mildly sweet wheat roti and a dry, slightly bitter bajra roti, most individuals would prefer the former, especially if they have never grown up eating the latter. Without cultural familiarity or emotional connection, they often get dismissed as bland. This, too, is reinforced by the collective perception of millets known as "Grain of the Poor." In rural areas, it is consumed as a last resort. In urban areas, it is sold at a much higher rate in organic stores.

Along with limited shelf life, increased cooking effort, and taste, the decline of millet usage in India is also influenced by organisational decisions, policy priorities, and market neglect. Figure 4.4 illustrates a map of these interconnected pain points. High-urgency issues like PDS exclusion, low urban awareness, and farmer disinterest sit right next to medium-urgency, but there are equally significant opportunities in reshaping cultural perceptions, reducing market hesitancy, and creating alignment among fragmented stakeholders. In urban markets, refined wheat, polished rice, and packaged bread sell fast. This also contributes to the decline in usage of millets.

After mapping the pain points in the millet ecosystem, it was evident that the millet revival required multi-layered interventions that target both systemic barriers and daily touchpoints in the consumer journey. The mapped out intervention points tackled a key bottleneck in a specific layer of the system, creating an integrated approach to millet resurgence. These interventions included:

1. Collaborating with quick-service restaurants.
2. Improving millet literacy and developing experiential stores.
3. Providing access to millet through public distribution ration shops.
4. Integrating millets into street food.
5. Providing policy support to millet farmers.

Together, these interventions formed a loop, starting from production, moving to distribution and awareness, and closing at consumption.

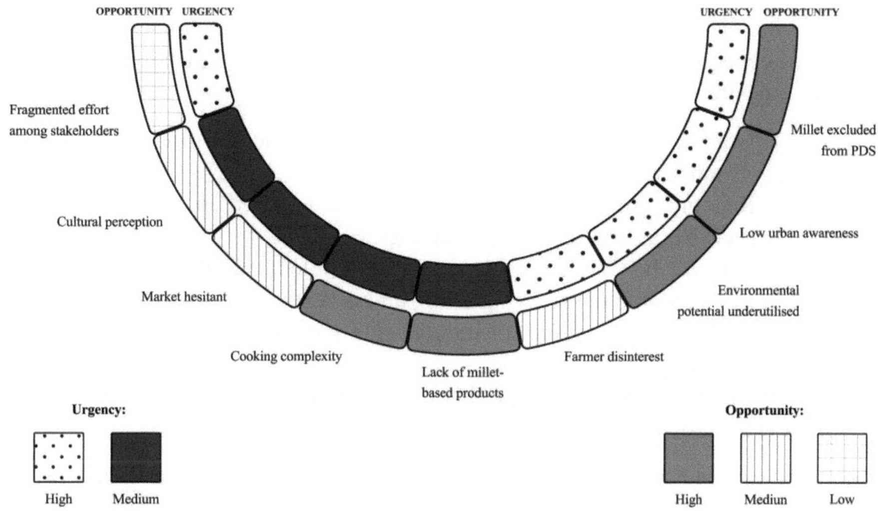

Fig. 4.4 Map of interconnected pain points

After addressing key systemic and user-level barriers of the five targeted interventions, the next step was to evaluate their real-world potential. In order to do this, a Value Proposition Framework was used. This framework evaluates each intervention across five criteria: Affordability, Innovation, Executability, Impact, Time Efficiency. Each criterion was measured on a scale of 1–10 based on both qualitative insights and feasibility analysis. The evaluation exercise was conducted to prioritise actionable, scalable solutions in the short and long term.

Based on this exercise, Intervention 2 stood out as the optimal solution that focuses on building awareness and bringing about behavioral change. It laid the groundwork for a millet-forward future, one that is inclusive, nutritious, and sustainable.

4.5 Envisioning the Interventions

The Indian consumers have been inadvertently avoiding millet due to limited recognition and lack of familiarity. Public perception of millets was identified as one of the strongest anchors for change and hence was considered as the leverage point required for the Intervention 2.

Once people get familiar with millets, how it can be consumed, its benefits, they would be more inclined towards adapting it as a part of their diet. This would then consequently push for a shift in the system. Changes in food culture would likely influence millet consumption patterns, potentially increasing supply and yielding benefits for farmers.

While the visible reasons for avoiding millets represent only a small part of the issue, the deeper challenges remain hidden beneath the surface, much like an iceberg. They are invisible yet critical:

- Lack of Awareness: Most people remain unaware of the nutritional benefits and diverse consumption methods of millets.
- Lack of Recognition: Millets rarely tend to be featured in mainstream food marketing, cooking shows, or restaurant menus.
- Don't Try, Don't Buy: A causal loop, where limited availability or display of millets leads to people never trying them, and as a result they never adopt them.
- Health Benefits: The incredible nutritional value of millets is under-communicated in daily discourse.
- Environmental Benefits: Despite being climate-resilient and water-efficient, millets are missing from sustainability conversations.

These underlying issues silently shape choices, culture, and access, and unless addressed, the surface-level solutions would not sustain. The key takeaway here was—"Solving the millet problem isn't just about easier recipes or tastier meals. It's about creating recognition, visibility, and systems that support change, both on the plate and in the mind.

Changing a food system goes beyond increasing supply or tweaking policy; it is largely about transforming human behavior. This transformation is not limited to promoting the benefits of millets, it lies in integrating a curated experience. People need to experience millets: taste them, cook with them, engage with them, and overcome the small struggles along the way, until they get accustomed to them. At its core, the gap between millets and modern diets isn't about availability; it's about the disconnect.

4.6 Envisioning the Interventions

In order to address this gap, an immersive, interactive solution was proposed—The Millet Story. It is a physical and digital platform designed to inform, engage, and inspire. The core concept of the solution is that for awareness to be effective, curiosity of the end user is important (Fig. 4.5).

The Millet Story would be an experiential store. A place where people can explore millet not as a health product, but as a culture. The layout is intentional. It is designed to be a space where people are able to learn about the varieties in millets, taste dishes from across the country, attend cooking workshops, and also observe how millet is grown.

There is also a proposed educational zone with interactive panels and storytelling walls that explain the history, health benefits, and cultural importance of millets. A workshop and event area where chefs, farmers, nutritionists, and storytellers come together. A tasting zone and a retail section, so the experience continues with the individuals even outside the store. It would be placed in an urban setting, next to

Fig. 4.5 The Millet Story—products and store layout

cafés, bookstores, or grocery hubs, in areas where people are already in the mindset to explore.

A strong digital identity would be vital for the successful implementation of the intervention. The social media handles would not be limited to marketing the products, but also tell stories of farmers involved in the millet ecosystem, share everyday millet recipes, collaborate with home cooks sharing their ancestral recipes, showcase behind-the-scenes prep, and extend the invitation for the community to join in. The goal is to reintroduce millet to the public. This platform would enable reintegration of millets among the public through workshops, farm modules, and everyday storytelling.

4.7 Conclusion

The current project started with rediscovering the wide world of millets using the proposed solution of a branded store. With the right exposure and a little momentum, a well-designed small shift in the system can spark a larger transformation. The proposed solution also aligns with the Sustainable Development Goals of Good Health & Well-being (Goal 03), Responsible Consumption & Production (Goal 12) and Climate Action (Goal 13). A human-centred design approach in systems focuses on creating experiences that invite participation rather than imposing change. The design solution sparks interest, while building trust, and allows change to grow, not as a mandate, but as a movement to support the growth of millet consumption.

References

1. Tiwari S (2020) MSP and public procurement. PRS Legislative Research. https://prsindia.org/theprsblog/msp-and-public-procurement
2. Pokharia A, Kharakwal J, Srivastava A (2014) Archaeobotanical evidence of millets in the Indian subcontinent with some observations on their role in the Indus civilization. J Archaeol Sci 42(1):442–455

3. Wordswork, SM (2022) Around India through its different breads, https://www.onmanorama.com/food/features/2022/03/13/india-through-different-breads.html
4. Explains (2022) Part 2: Important food crops (Rice, wheat, maize, millets, pulses and barley) and horticultural crops. CivilsDaily. https://www.civilsdaily.com/important-food-crops-rice-wheat-maize-millets-pulses-geographical-conditions-producing-areas-important-varieties-horticulture-fruits-vegetables/
5. Largest Millets producing states in India 2025—Top 10 list. Testbook. https://testbook.com/static-gk/largest-producer-of-millets-in-india
6. National Institute of Diabetes and Digestive and Kidney Diseases. Your digestive systems & how it works? https://www.niddk.nih.gov/health-information/digestive-diseases/digestive-system-how-it-works
7. Yacoubian J, Grigoryan A, Grigoryan A (2023) Rice vs. Millet: health impact and nutrition comparison. https://foodstruct.com/nutrition-comparison-text/rice-vs-millet-raw
8. Shah P, Mehta N, Shah S (2023) Exploring the factors that drive millet consumption: insights from regular and occasional consumers. J Retail Consum Serv 76:103598
9. Millets: The Future Food. India Science, Technology & Innovation (ISTI) Portal. https://www.indiascienceandtechnology.gov.in/listingpage/millets-future-food
10. India Data Sights (2023) Millet cultivation in India: history and trends. India Development Review. https://idronline.org/article/agriculture/millet-cultivation-history-and-trends/?
11. Bandyopadhyay U, Patnaik A (2025) From green revolution to millet revolution: understanding India's transition through agri-food policies. Food Cult Soc 29:220–240. https://doi.org/10.1080/15528014.2025.2485505
12. Sruthi NU, Rao PS (2021) Effect of processing on storage stability of millet flour: a review. Trends Food Sci Technol 112:58–74. https://doi.org/10.1016/j.tifs.2021.03.043
13. Menon A (2023) The return of the forgotten millet—the Locavore. The Locavore. https://thelocavore.in/2023/03/02/the-return-of-the-forgotten-millet/
14. Meal MM (2024) Glycemic index of millets: a diabetic-friendly superfood. millet magic meal. https://www.milletmaagicmeal.in/blogs/blogs/glycemic-index-of-millets?
15. Sreekala ADS, Anbukkani P, Singh A, Rao BD, Jha GK (2022) Millet production and consumption in India: where do we stand and where do we go? Natl Acad Sci Lett 46:65–70. https://doi.org/10.1007/s40009-022-01164-0
16. Singh G (2023) Food of the future: the business and economy of super grains. https://economictimes.indiatimes.com/industry/cons-products/food/food-of-the-future-the-business-and-economy-of-super-grains/articleshow/99336226.cms?from=mdr

Chapter 5
Mushrooms and Health Remediation: Steps Towards Air Purification

Aditi Sharma, Saili Mandavkar, and Samruddhi Khemkar

Abstract The Indian mushroom market is flourishing rapidly due to its economic value, environmental benefits, and role in improving nutrition. Commercial mushroom farming in India began in the late 1960s after its advantages were highlighted by the Development of Mushroom Cultivation in Himachal Pradesh project. This project aims to understand the current oyster mushroom market, analyze the factors involved in the production, distribution, and support, and explore the potential of mushrooms as a sustainable resource for innovative applications. The mushroom-related markets and the stakeholders were categorized into primary, secondary, and tertiary categories. The cultivation process was analyzed, which included the substrate preparation, spawning, incubation, fruiting, harvesting, and post-harvest management. This led to the identification of key areas for improvement and innovation. Due to its versatility and increasing popularity, Oyster Mushrooms (Pleurotus ostreatus) were selected as the focus of the project. The study further examined the industrial and commercial applications of oyster mushrooms, focusing on mycoremediation. Mycoremediation is a process that uses fungi to clean up environmental pollutants. This project also addressed the challenges and limitations in commercializing mycoremediation. The regenerative nature of mushrooms, along with their low carbon footprint, natural purifying properties, and rapid growth cycle, makes them a valuable resource for sustainability and the circular economy. A mycelium-based air purifier was proposed to integrate sustainability into everyday spaces and demonstrate the practical applications of fungi in environmental innovation as an outcome of the project.

Keywords Oyster mushrooms · Mycoremediation · Circular economy · Bio-inspired design

---

A. Sharma · S. Mandavkar (✉) · S. Khemkar
Indian Institute of Technology Kanpur, Kanpur, India
e-mail: slmandavkar7@gmail.com; sailim24@iitk.ac.in

V. Kant (ed.), *Human Centred Design for Systems*, Springer Series in Design and Innovation 68, https://doi.org/10.1007/978-3-032-30376-9_5

5.1 Introduction

A mushroom is a fleshy, spore-bearing fruiting body of a fungus, typically growing above ground, on soil, or on its food source. They come in a wide range of sizes and colors, and can be found almost anywhere, from forests to your local grocery store. They grow from a network of tiny, branching fibers called mycelium. This mycelium can cover vast areas, breaking down organic matter and recycling nutrients. When conditions are right, the mycelium produces small bumps called primordia, which eventually develop into mature mushrooms. They play a vital role in the ecosystem as key decomposers, helping to break down dead organic matter and recycle nutrients. Some fungi form symbiotic relationships with plants, facilitating communication and nutrient exchange through vast networks that can be called the 'Wood, Wide Web'. The Indian mushroom market is growing rapidly due to its economic potential, sustainability benefits, and contribution to nutritional security.

5.2 Framing the System

Mushrooms are beneficial for the environment due to their capacity to mitigate pollution. Commercially, however, this has not been leveraged to its full extent. Cultivation of mushrooms, too, has not been a widespread activity due to its limited consumption. The Development of Mushroom Cultivation in Himachal Pradesh project [1] highlighted the benefits of commercial mushroom farming, such as employment generation, improved income prospects, and increased crop variety. This laid the foundation for what would become a thriving industry. Globalization also contributed as a factor that helped shift the perception of mushrooms as a valuable protein source and a healthy substitute for meat.

The project began with an exploration of the mushroom species, including Button, Shiitake, Oyster, Portobello, Enoki, Lion's Mane, Lingzhi, and Truffle. Among the various types of mushrooms, Oyster Mushrooms (Fig. 5.1) were selected for further research due to their versatile features and substantial benefits.

Oyster Mushroom cultivation process involves several key steps: Preparing the substrate, spawning, incubation, fruiting, harvesting, and post-harvest handling [2]. The first step is arranging the substrate. Oyster mushrooms can grow on a variety of materials, including straw, sawdust, coffee grounds, and even cardboard. The substrate must be pasteurized to eliminate unwanted microorganisms. This can be done through heat pasteurization, lime bath treatment, or peroxide treatment. Once the substrate is prepared, it's time for spawning. Mushroom spawn, which is essentially the mycelium of the mushroom, is mixed into the pasteurized substrate.

This mixture is then placed in a clean environment to prevent contamination. The next stage is incubation. The substrate and spawn mixture is incubated in a warm, dark place with high humidity. This allows the mycelium to colonize the substrate fully. The ideal temperature for incubation ranges from 22 to 28 °C, and

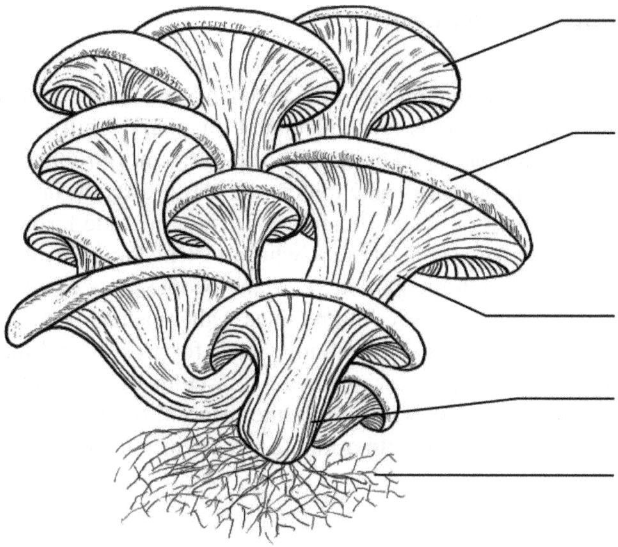

Fig. 5.1 Parts of oyster mushrooms

humidity should be maintained at 65–75% in complete darkness and an optimal carbon dioxide level in the range of 5,000–10,000 ppm. The fruiting initiation phase triggers mushroom growth by adjusting environmental conditions [2]. Lower temperatures of 15–22 °C, increased fresh air exchange, indirect light of 500–1000 lx, and high humidity (95%) encourage primordia formation within 3–5 days. Proper control of these factors ensures successful fruiting and a healthy yield. After the substrate is fully colonized, it's time for the fruiting phase. The environment is adjusted to trigger mushroom growth. This includes providing indirect light, maintaining high humidity, and ensuring adequate airflow. The temperature is reduced to around 16–22 °C with the humidity at 85–95% and a reduction of CO_2 levels to be below 1000 ppm, in either indirect sunlight or artificial lighting at 500–1000 lx, in order to stimulate fruiting. Once the mushrooms have grown to the desired size, they are ready to be harvested. Oyster mushrooms are typically harvested when the caps begin to uncurl and flatten. They can be gently twisted and pulled or cut at the base of the stem. After harvesting, oyster mushrooms are highly perishable and must be stored in cool conditions to extend their shelf life. They can be stored in the refrigerator at around 4 °C for several days. For longer storage, freezing or dehydrating methods are used.

5.3 Comprehending the System

The trends in the oyster mushroom market are driven by consumer demand for healthier and more sustainable options. To understand this market fully, the roles of primary, secondary, and tertiary stakeholders of the oyster mushroom industry were identified (Fig. 5.2) [2]. Primary stakeholders are those directly involved in the production and supply chain of oyster mushrooms. These include farmers who cultivate the mushrooms, workers and laborers who assist in the cultivation and harvesting processes, along with home gardeners. Many people grow oyster mushrooms for personal consumption or lesser volume sales. Other primary stakeholders include spawn producers who supply the mycelium needed for growth, and substrate suppliers who provide materials like straw or sawdust for cultivation. Additionally, there are waste management workers who handle the disposal of spent substrates, and odor management facilitators who help mitigate any environmental impacts.

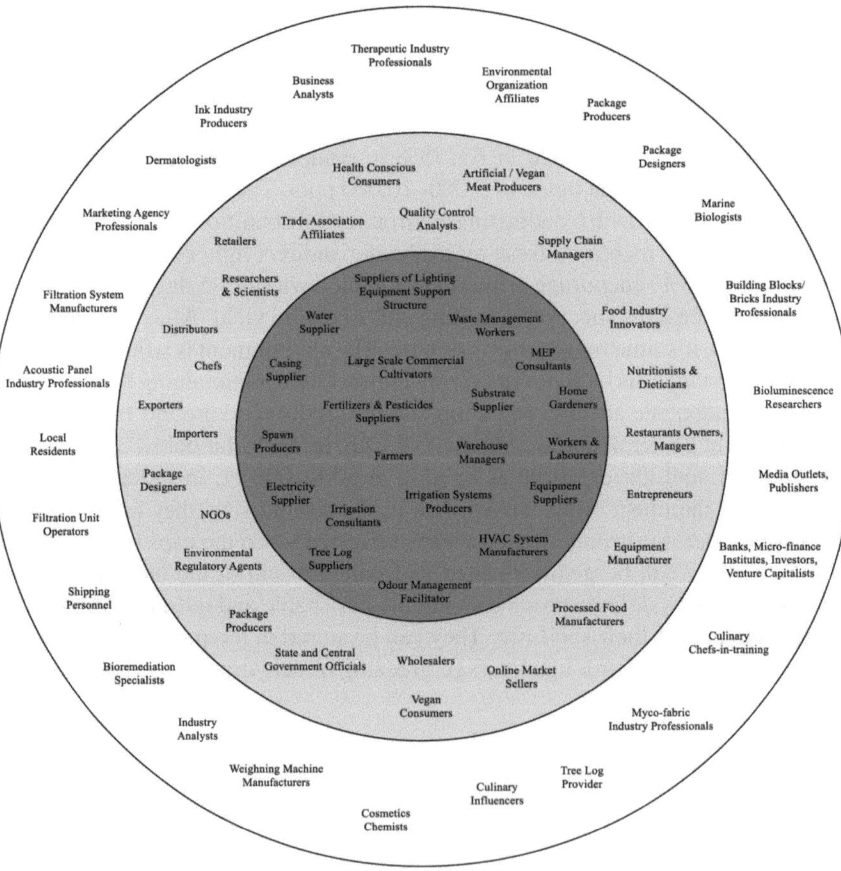

Fig. 5.2 Oyster mushroom stakeholders

Secondary stakeholders are those who influence the market indirectly. These include trade association affiliates who advocate for the industry, food industry innovators who develop new products using oyster mushrooms, health-conscious consumers who drive demand for nutritious and sustainable food options, package producers who design and supply packaging materials for mushroom products, amongst others. Tertiary stakeholders are those who benefit from or contribute to the broader ecosystem surrounding oyster mushrooms. These include culinary chefs who learn to incorporate oyster mushrooms into recipes, bioremediation specialists who utilize mushrooms for environmental cleanup, myco-fabric industry professionals who develop sustainable textiles from mycelium, and therapeutic industry professionals who explore the medicinal properties of oyster mushrooms, to name a few.

The gigamap illustrates how these stakeholders interact and influence each other. For example, farmers rely on spawn producers for quality mycelium, while health-conscious consumers drive demand for organic products. This interplay shapes market dynamics from supply chain efficiency to product innovation and environmental sustainability. With increasing interest in plant-based diets and sustainable practices, this industry is poised for significant growth. When it comes to industrial applications, pharmaceuticals and nutraceuticals prefer oyster mushrooms for their anti-cancer, anti-inflammatory, and cholesterol-lowering properties [3]. These properties make them a promising ingredient in medicinal compounds. They are also used in making nutritional supplements, as they are rich in vitamins, minerals, antioxidants and bioactive compounds. In the cosmetics industry, oyster mushrooms are used to develop natural skincare products. Their antioxidant and anti-inflammatory properties help combat oxidative stress and inflammation, promoting healthier skin. Products like creams and serums utilize oyster mushroom extracts to enhance skin health and reduce signs of aging. Table 5.1 shows the industry applications of oyster mushrooms [3].

In the textile industry, oyster mushrooms are used to create innovative materials like mycofabric and faux leather. The mycelium of these mushrooms can bind plant fibers together, forming a durable and sustainable textile. These materials are lightweight, flexible, and have insulating properties, making them suitable for various applications. Similarly, the Golden and Pink Oyster mushrooms can be used to extract completely natural dyes. In agriculture, spent oyster mushroom substrates can be repurposed as organic manure. This substrate is rich in nutrients and can enhance soil fertility, supporting sustainable farming practices. Additionally, oyster mushrooms can be used as feed for cattle and other animals, providing them with essential nutrients. Biomaterials derived from mycelium are being explored for construction and other industrial applications. These materials are made into bricks or panels, which are lightweight, fire-resistant, and have acoustic properties. They are biodegradable and sustainable, offering a promising alternative to traditional building materials.

Oyster mushrooms possess a fascinating array of properties that make them not only a culinary delight but also a powerful tool in environmental cleanup. Following are biological, physical, chemical, and cultural properties of oyster mushrooms:

Table 5.1 Industrial applications of oyster mushrooms [3]

Industry	Applications
Food industry	Culinary use, Meat substitute, Processed food and Snacks
Pharmaceutical and Nutraceutical	Medicinal compound, Anti-cancer, Anti-inflammatory, Cholesterol-lowering property, Nutritional supplements
Agriculture	Bioremediation—absorbs heavy metals and pollutants from soil and water, Composting / Animal feed
Textile	Natural dyes, Faux leather
Research and development	Scientific research
Cosmetics	Haircare, Skincare
Aquaculture	Fish feed supplement
Paper production	Mycelium based paper
Biomaterial	Bricks, Acoustic panels
Biofuel	Residual substrate used in bio-ethanol production
Gardening	Diy Grow-At-Home mushroom kits
Art and design	Furniture, Sculpture
Ink production	–

Biologically, they are anti-inflammatory and exhibit antibacterial effects against both Gram-positive and Gram-negative bacteria. Certain compounds present in them inhibit cancer cell growth and help cancer cells die. Its mycelium actively metabolizes and decomposes pollutants through enzymatic action. Physically, oyster mushrooms are typically fan-shaped or oyster-shaped, with diameters ranging from 5–15 cm or more, depending on the species and growth conditions. They are commonly found in shades of white, gray, brown, or yellow, depending on the species and maturity level, and are characterized by their extensive mycelium network. This network can cover large areas, allowing the mushrooms to interact with pollutants more effectively, as it breaks down organic matter, and absorbs the broken down pollutants from soil and water. The dense, fibrous structure of the mycelium acts as a natural filter, trapping particulate matter and enhancing airflow, while capturing dust and other airborne contaminants.

Chemically, oyster mushrooms are rich in essential minerals such as copper, selenium, potassium, and phosphorus, and contain essential amino acids, making them a good source of protein. Its mycelial network secretes enzymes which help in breaking down organic pollutants in soil, air, and water. These enzymes, such as laccases and manganese peroxidases, are key in degrading complex pollutants like aromatic hydrocarbons and pesticides. Additionally, the property of biosorption enables oyster mushrooms to effectively absorb and accumulate heavy metals and pollutants from contaminated environments, be it soil, water, or air. These properties make them excellent for bioremediation. Thus, by using oyster mushrooms, contaminated environments can be cleaned up naturally and sustainably. Culturally, oyster mushrooms have been valued for centuries in various cuisines and traditional medicines.

Bioremediation is an environmentally friendly process that uses living organisms to neutralize or remove contaminants from polluted environments. This method harnesses the natural metabolic processes of microorganisms, plants, or fungi to degrade hazardous substances into less toxic or non-toxic forms, thereby restoring ecological balance. A specialized form of bioremediation is mycoremediation, which utilizes fungi, particularly their mycelium networks, to decompose environmental pollutants [4]. The mycelial network excretes enzymes and acids, which actively seek out lignin and cellulose. These compounds consist of long chains of carbon and hydrogen. Interestingly, many organic pollutants, such as petroleum products, pesticides, and herbicides, are composed of similar carbon-hydrogen structures. This biochemical similarity allows fungi to potentially break down these pollutants in a similar manner to how they decompose natural plant material.

Paul Stamets, a well-known mycologist and an advocate of medicinal fungi and mycoremediation, carried out experimental work on mycobooms [5]. Mycobooms are oyster mushroom-inoculated materials designed for environmental remediation. They consist of 100-percent biodegradable hemp socks, filled with straw, mixed with germinated mushroom spores. These mycobooms absorb oil above the waterline while the fungi inside break down the straw, producing CO_2, water, acid, and enzymes. According to Paul Stamets, the mycelium's fragrance attracts insects and fish, creating a beneficial domino effect in the ecosystem. Mycobooms act as both a barrier to prevent new oil from spreading and an active remediation tool, releasing enzymes to break down existing oil in marshland habitats.

The mycoremediation properties of fungi have been known for over a century, dating back to World War I, yet their commercial application remains limited. Despite extensive research, practical implementation in large-scale environmental cleanup has been slow. One of the most promising fungal species in this field is the Oyster Mushroom, which has demonstrated remarkable efficiency in breaking down pollutants, including hydrocarbons and heavy metals. However, while significant studies have validated its potential, challenges in scalability, regulatory approvals, and economic feasibility have hindered its widespread adoption. While oyster mushrooms and other fungi have demonstrated remarkable potential in mycoremediation, their real-world application is influenced by multiple factors beyond biological capabilities.

Implementing mycoremediation at scale required collaboration among diverse stakeholders (Fig. 5.3), each playing a vital role in research, development, execution, and regulation. These stakeholders have been extracted by sifting through large numbers of primary and secondary literature. At the core are mycologists, who study fungi and their interactions with contaminants. They focus on identifying fungal species capable of biodegradation, biosorption, and bioconversion. Mycologists collaborate with mushroom cultivators and suppliers to test new fungal strains and optimize their growth for remediation purposes. Cultivators ensure consistent strain performance by implementing rigorous quality control measures and conducting training sessions to enhance cultivation techniques.

Substrate producers and suppliers provide agricultural and industrial waste as growth mediums for fungi. These substrates support fungal growth while addressing

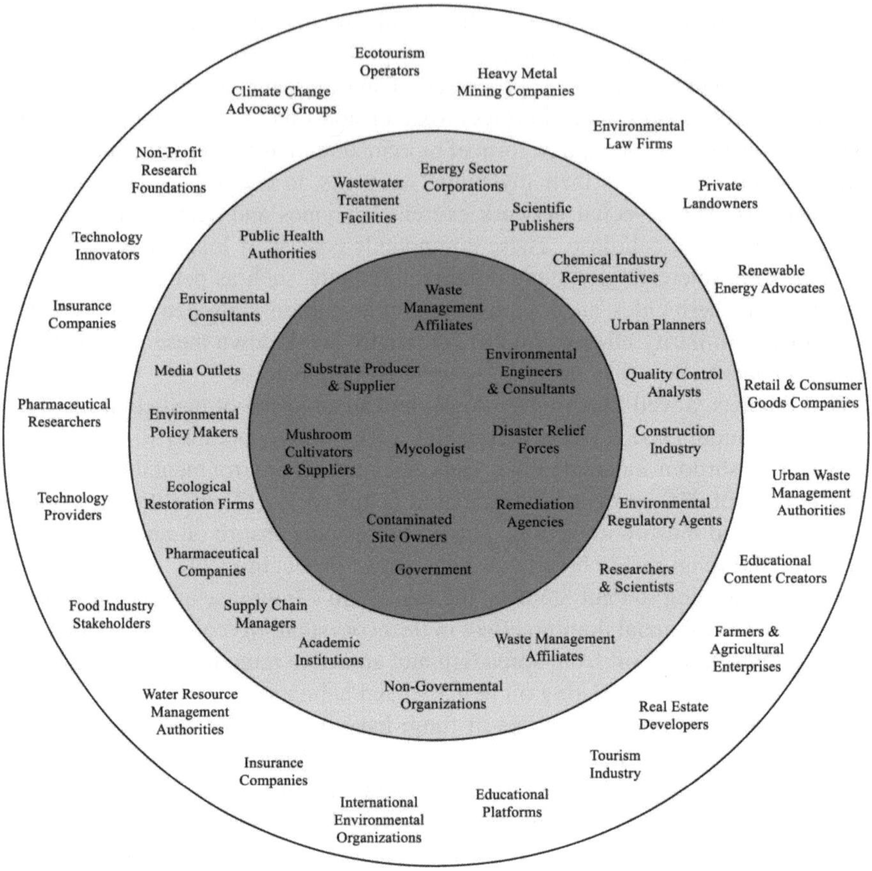

Fig. 5.3 Mycoremediation mushroom stakeholders

waste management challenges [4–6]. Environmental engineers and consultants work closely with both substrate providers and cultivators to explore the sustainability of by-products, ensuring the process is environmentally sound. Remediation agencies play a pivotal role in translating laboratory findings into real-world applications. They analyze how fungi interact with specific contaminants and substrates, employing strategies to implement mycoremediation at the contaminated sites. These agencies often collaborate with government bodies that facilitate partnerships between public and private sectors, develop regulatory frameworks, and promote awareness about mycoremediation's benefits [6–8].

Waste management affiliates contribute by identifying suitable waste sources for fungal remediation. They conduct human health risk assessments related to the contaminated sites. Similarly, contaminated site owners work with remediation agencies to evaluate the potential genotoxicity or mutagenicity of fungi grown on polluted sites. In emergency scenarios, disaster relief forces collaborate with mycologists and remediation agencies to deploy fungi for rapid environmental recovery [5, 8]. The government also supports these efforts by incentivizing research and implementation. Finally, there is growing interest in combining remediation with food production by exploring the potential of cultivating edible mushrooms on non-toxic waste sources [4, 5]. These stakeholders are interconnected through research, resource sharing, policy development, and practical implementation, collectively advancing the field of mycoremediation while addressing environmental challenges sustainably. After narrowing down the primary stakeholders, the next phase involved understanding the mechanism behind the mycoremediation process. The process is a synergy of biological and chemical processes that enable fungi to degrade, absorb, or transform harmful pollutants into less toxic forms. Mycoremediation relies on fungi, particularly their mycelial networks, to decontaminate polluted environments [4, 5].

Here, the first step involves selecting the right fungal species. Not all fungi are suitable for remediation; decomposer fungi, such as oyster mushrooms, are often chosen for their robust enzyme systems and adaptability. These fungi are screened based on their tolerance to specific pollutants and environmental conditions. The second step involves recognition of the contaminant. Once introduced into a polluted site, the fungal mycelium detects contaminants in its surroundings. This recognition is facilitated by the fungi's ability to sense environmental changes and respond by producing specialized enzymes or acids tailored to break down or interact with the pollutants [5, 8]. The next step involves the uptake of the contaminant, where the mycelial network, which can extend over vast areas, penetrates contaminated soil or water. Through its extensive surface area, it absorbs pollutants directly into its biomass. This uptake is critical for processes like biosorption and bioaccumulation [7, 9].

In the fourth step of enzymatic degradation, the fungi produce non-specific extracellular enzymes such as laccases, peroxidases, and cellulases. These enzymes break down complex organic pollutants like hydrocarbons, pesticides, and dyes into simpler, non-toxic molecules such as carbon dioxide and water. For example, oyster mushrooms have been shown to degrade petroleum hydrocarbons effectively [4, 5, 8]. The fifth step involves bioaccumulation and transformation. Here, some pollutants, particularly heavy metals, are stored within the fungal biomass through bioaccumulation. Others undergo biotransformation that involves chemical modifications that convert them into less harmful forms. Fungi can even break metals using organic acids like oxalic acid [4, 5]. Lastly, the sixth step involves nutrient cycling and ecosystem recovery, where the fungi degrade pollutants and release the nutrients back into the environment, promoting ecosystem recovery. Additionally, their activity fosters microbial diversity by creating conditions suitable for bacterial growth and plant regeneration [4, 10].

Biodegradation, biosorption, and bioconversion are the three primary decontamination methods that fungi employ to remediate polluted environments [4]. Biodegradation is the enzymatic breakdown of complex organic pollutants into simpler, less harmful substances. Fungi like 'Pleurotus Ostreatus' and 'Trametes Versicolor' produce extracellular enzymes such as laccases and peroxidases, which degrade persistent pollutants like hydrocarbons, pesticides, and pharmaceuticals. This method is highly effective for organic contaminants and can be applied both in on-site and in controlled environments. However, biodegradation requires careful optimization of environmental factors like pH, temperature, and nutrient availability to ensure maximum efficiency [4, 5]. Biosorption leverages the fungal biomass, either living or dead, to adsorb heavy metals and other pollutants from contaminated water or soil. Components of the fungal cell wall, such as chitin and glucans, bind to metals like lead, mercury, and uranium through ion exchange or complexation. Fungi like 'Fomes Fomentarius' and 'Ganoderma Lucidum' are particularly effective for this process. Unlike biodegradation, biosorption does not rely on metabolic activity, making it a cost-effective option for industrial pollutants. Dead fungal biomass is often preferred due to its stability under harsh conditions. Lastly, bioconversion focuses on transforming waste materials into valuable products through fungal metabolic processes. For example, agricultural waste can be converted into edible mushrooms or biofuels using species like 'Pleurotus Ostreatus' and 'Agaricus Bisporus' [4, 5]. This method not only addresses waste management challenges but also contributes to resource recovery by producing usable by-products. It typically requires controlled conditions to ensure product safety and effectiveness.

The success of mycoremediation is influenced by several interconnected factors. One of the most critical aspects is the selection of fungal species. Different fungal species exhibit varying enzymatic capabilities to break down hydrocarbons, heavy metals, and other toxins; hence, identifying fungi compatible with the specific pollutants in a contaminated site is essential [5]. For instance, oyster mushrooms have demonstrated significant efficacy in degrading petroleum hydrocarbons due to their ability to produce enzymes like laccases and peroxidases. The type and concentration of contaminants also play a pivotal role. Sites with high levels of complex pollutants may require additional treatment techniques or synergistic approaches involving co-cultivation with bacteria that enhance fungi's degradation capabilities. Environmental conditions such as temperature, pH, moisture content, oxygen availability, and soil porosity are equally important. Fungi thrive within specific temperature ranges (generally 25–30 °C) and pH levels conducive to enzymatic activity. Soil porosity and water availability further influence fungal colonization [1, 7].

Nutrient availability is another critical factor. Fungi require carbon, nitrogen, phosphorus, and mediators like vanillin to sustain their growth and enzymatic activity. These nutrients not only support fungal metabolism but also enhance their ability to break down contaminants effectively [2, 5]. Additionally, co-cultivation and synergistic relationships between fungi and other microorganisms can amplify remediation outcomes by creating symbiotic systems that tackle pollutants more efficiently. Site-specific factors, such as soil composition, vegetation, past land use patterns, microbial competition, and accessibility of contaminants to fungi, also determine the success of mycoremediation efforts [4, 7]. Contaminant bioavailability can be improved by using surface-active agents that help overcome barriers during remediation processes. For volatile compounds in contaminated soils, vapor-off-gas controls may be required to manage emissions effectively. Ultimately, the interplay between these factors underscores the complexity of designing successful mycoremediation systems. By tailoring fungal species selection and optimizing environmental conditions, nutrient supply, and site-specific strategies, mycoremediation can be a powerful tool for ecological restoration [5].

The commercialization of mycoremediation faces significant challenges and barriers despite its promising potential as a sustainable solution for environmental restoration. One major obstacle is the slow pace of fungal colonization and pollutant degradation compared to conventional remediation techniques. Fungi require time to establish themselves in contaminated environments, which can delay the remediation process and make it less appealing for industries seeking faster solutions. Furthermore, fungi are living organisms whose performance is inherently unpredictable, as they are influenced by environmental variables such as temperature, pH, and moisture levels. This lack of control and predictability often necessitates site-specific strategies, which can be resource-intensive and difficult to scale [4, 5, 8]. Another limitation is the restricted range of contaminants that fungi can effectively degrade. While fungi excel at breaking down organic pollutants like hydrocarbons, they often lack adequate mechanisms to detoxify or remove heavy metals and other inorganic toxins. This limitation reduces their applicability in diverse contamination scenarios, further hindering widespread adoption. Additionally, mycoremediation remains largely confined to lab or greenhouse settings, with limited field-scale evaluations available [5, 7]. This gap between experimental success and practical implementation raises concerns about its reliability in real-world conditions.

Economic and infrastructural barriers also impede commercialization efforts. The lack of standardized protocols for mycoremediation complicates its integration into existing waste management systems [7, 8]. Testing costs, scaling difficulties, and the need for specialized infrastructure add financial burdens that deter investment from industries and governments alike. Regulatory barriers further exacerbate these challenges, as there is often insufficient certification or legal support for deploying fungal-based remediation methods. Lastly, the market competition and short-term economic goals discourage stakeholders from pursuing mycoremediation as a viable alternative. Investors perceive higher risks associated with unproven technologies, while waste management agencies prioritize cost-effective solutions over long-term sustainability. These factors collectively contribute to the underutilization of mycoremediation despite decades of research. Bridging these gaps will require collaborative efforts across research institutions, governments, investors, and industries to establish standardized practices, improve scalability, and demonstrate field-level efficacy to unlock its commercial potential. Therefore, the key issue is to recognize the number of interactions occurring at various levels of the system that evolve over several spatial and temporal scales.

Mycoremediation presents an innovative and eco-friendly approach to pollution control, but it faces several challenges and limitations that hinder its widespread application. One major concern is its limited commercial application with minimal progress in large-scale deployment, particularly for petroleum-contaminated soils. Additionally, the slow process of fungal colonization and establishment makes it less efficient than conventional remediation methods, which often yield quicker results [5, 8]. Another significant challenge is the limited range of contaminants that fungi can effectively neutralize [6]. While they work well for organic pollutants, they lack efficient mechanisms to detoxify heavy metals. Furthermore, mycoremediation suffers from a lack of control and predictability as fungi are living organisms highly sensitive to environmental factors like temperature, moisture, and competition from other microbes [2, 5].

The need for site-specific considerations adds another layer of complexity, thus requiring tailored approaches for different locations and contamination types. Most research is still confined to controlled or greenhouse environments, making its real-world effectiveness uncertain. To fully harness the potential of mycoremediation, further research, technological advancements, and strategic implementation are necessary to address these challenges [5]. India faces a range of environmental challenges that make mycoremediation a promising solution for addressing industrial waste, heavy metal pollution, agricultural runoff, and oil spills [6]. Industrial waste, especially from sectors like textiles, pharmaceuticals, and paper-pulp, often contains complex organic pollutants and heavy metals. Fungi, with their enzymes and metal-binding proteins, can degrade organic toxins and bioleach heavy metals such as lead and mercury from contaminated soils and wastewater [4, 5].

Mycoremediation offers an eco-friendly alternative to conventional methods by breaking down these persistent chemicals into less harmful compounds, thereby restoring soil health [10]. Oil spills and marine pollution along India's extensive coastline further highlight the need for innovative solutions. Fungi such as 'Pleurotus Ostreatus' have shown potential in degrading hydrocarbons present in crude oil, making them suitable for remediating oil-contaminated beaches or mangroves in regions like Kerala or the Sundarbans. Additionally, surface water bodies like the Ganges are often polluted with industrial effluents and agricultural runoff. Mycoremediation could offer a cost-effective way to reduce pollution while also dealing with problems caused by nutrient-rich waste [5, 6]. By leveraging its vast fungal biodiversity and tailoring site-specific strategies, mycoremediation presents the promise of integration into environmental management practices [6, 7].

5.4 Envisioning the Interventions

With a deep understanding of the system, the next step was to identify challenges within the chosen mushroom application and propose strategic interventions. By leveraging insights from the background study, a design solution that enhances efficiency and sustainability within the ecosystem was developed. This required the use of a ZIP analysis. Drawing from the industrial applications of the oyster mushrooms, the initial focus was on mycoremediation. After analyzing the potential benefits and the challenges, the barriers for its commercialisation were discovered. Another intervention focused on using the fungal biomass generated after the mycoremediation process. As mentioned earlier, different methods of decontamination were explored; of these methods, only bioconversion had the byproduct of edible non-toxic mushrooms [4]. The concerns about the other methods of producing mushrooms remained.

How could the rest of the methods be repurposed in a way that avoids any potential revenge effects? Here, converting residual biomass into myco-textile and building materials showed promise.

This led us to explore Myco-booms. Myco-booms are a type of bioremediation tool that utilizes fungi, particularly oyster mushroom mycelium, to clean up environmental pollutants [8]. They are designed to absorb and break down contaminants such as oil spills. Myco-booms are typically made by colonizing straw or other organic materials with oyster mushroom mycelium. This mycelium can absorb oil and other pollutants, allowing for simultaneous absorption and degradation of contaminants. While myco-booms show great promise, there are challenges to their widespread adoption. These include optimizing the growth conditions for mycelium, scaling up production, and ensuring consistent efficacy across different types of pollutants [5, 11, 12]. Experimenting with different forms of Myco-booms to treat heavy metal poisoning in different topographies of India turned out to be the intervention point. Along with this, the use of decontamination methods to tackle indoor air pollution by leveraging the filtration and purification properties of the mycelium network was also considered.

These intervention points were mapped out on a Cost-Time graph as shown in Fig. 5.4 The intervention with the highest impact and the least effort required was selected. The fifth intervention of tackling air pollution was selected as the leverage point.

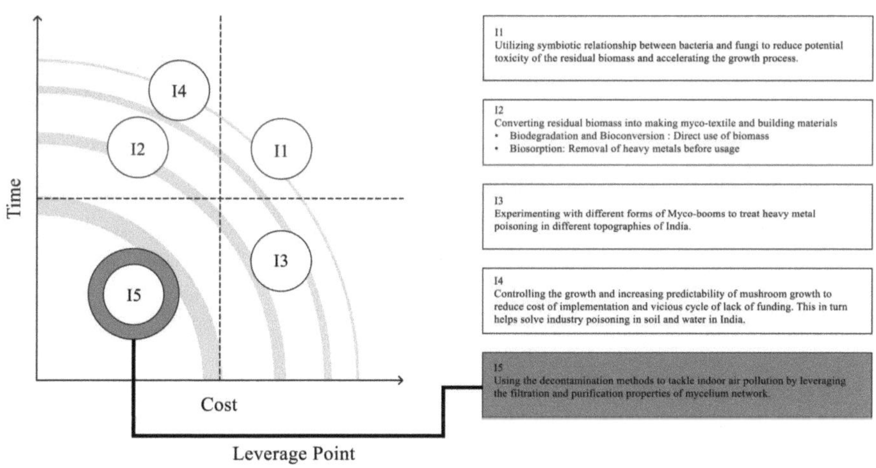

Fig. 5.4 Leverage points

5.5 Formulating the Interventions

Building on our identified leverage point, a sustainable, bio-inspired air purifier that harnesses the natural properties of mycelium for air purification, leveraging the principles of mycoremediation, was developed as the solution. Mycelium's natural filtration properties offer a sustainable way to combat indoor air pollution [13]. This intervention explored integrating mycelium-based systems into air filtration to reduce pollutants and improve air quality. With its regenerative nature, mycelium presented an eco-friendly solution for cleaner indoor environments (Fig. 5.5).

The air purifier construction includes a base housing the battery, battery module, valve actuators, and other electronic components, then comes a pre-filter with a suction fan on top to create a negative air pressure leading to air influx. Above the fan is the mycelium filter chamber with iris valve enclosures on either end to contain the air to be treated. A suspended, removable rod with a base of mycelium-inoculated substrate works as a natural HEPA filter. The size of the pores is to be 10–50 μm to maintain an average pulsating air flow rate. The top iris valve closes on this rod. The air then passes through an activated charcoal filter to get rid of the odour. Another fan on top creates pressure for the air outlet.

Fig. 5.5 Mechanism of the air purifier

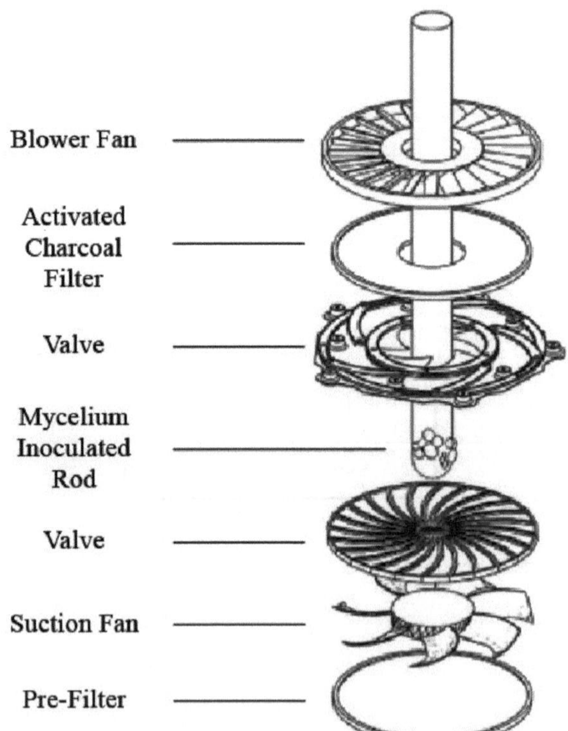

The form of the air filter draws inspiration from the organic morphology of oyster mushrooms, blending aesthetics and functionality through biomimetic principles. The design integrates natural curves and porous structures, mirroring mushrooms' gills and rhizomorphic growth patterns, to optimize airflow dynamics and user interaction. The form consists of a Bulbous Base, which is the lower section of the filter. Its rounded form is reminiscent of an oyster mushroom's cap. This part accommodates a mycelium filter chamber with a capacity of 48–50 cubic meters per hectare, ensuring sufficient air volume for effective purification. The expanded base enhances particulate matter retention by reducing airflow velocity, allowing mycelium networks to trap PM2.5 and PM10 particles [13]. A tapered upper section of the filter acts as a pressure regulator, accelerating airflow. This mimics the natural constriction found in mushroom stipes. The narrowed neck reduces aerodynamic resistance, improving energy efficiency while maintaining a consistent outflow of purified air.

The materials used in the filter include a transparent bioplastic chamber, which enables users to observe mycelial growth and particulate accumulation. This transparency fosters engagement, allowing users to monitor filter health and microbial activity without disassembly. A suspended, detachable rod houses mycelium-inoculated substrate, designed for easy harvesting. The rod's porous structure maximizes surface area for pollutant capture, while enabling airflow rates of 22–24 cubic meters per hectare [13]. The users can remove the rod to harvest mature mushrooms. The filter's design incorporates aesthetic and ergonomic considerations inspired by the oyster mushroom. Its organic curves mimic the mushroom's gently wavy cap, which helps reduce visual bulk and allows it to blend smoothly with indoor environments. Additionally, the base features small perforations that resemble the network of fungal hyphae, improving air intake while preventing clogging. These pores help maintain steady airflow, striking a balance between effective filtration and energy efficiency. The bioplastic enclosure and mycelium filter are fully compostable, contrasting with synthetic HEPA filters that contribute to landfill waste [9]. This biomimetic design not only enhances air quality but also transforms purification systems into living, sustainable artifacts, thus proving that nature's solutions often surpass human-engineered alternatives.

5.6 Enabling the Transition

The proposed product, "Aeolix" (Fig. 5.6) is inspired by Aeolus, the ruler of the winds, suggesting control over airflow and purification. This brings the process of mycroremediation into homes. HEPA filters are widely recognized for their use in air purifiers and effectiveness in capturing airborne particles, but they also come with several limitations and issues. HEPA filters are non-biodegradable and require regular replacement, contributing to landfill waste. The pollutants they capture remain on the filter, potentially causing environmental harm upon disposal [13].

Some HEPA air purifiers emit ozone as a byproduct, which can irritate respiratory systems and worsen indoor air quality. On the other hand, High-quality HEPA filters

Fig. 5.6 Proposed Product, "Aeolix"

are expensive, and regular replacements add to the overall cost of maintaining an air purification system. The selected intervention point is to replace the HEPA filter with something more natural, like the filtration of air by the mycelial network. The primary objective was to align the existing system with circular economy principles, ensuring minimal environmental impact while significantly enhancing indoor air quality. By integrating mycelium-based filtration, the aim was to reduce waste generation and reliance on non-renewable materials, creating a more sustainable alternative to traditional air purification systems. A conventional air purifier typically employs a three-stage filtration system. The prefilter, often a washable fabric, captures larger particles like dust and hair, extending the lifespan of subsequent filters. The HEPA filter targets fine particulate matter, trapping 99% of particles, including allergens, bacteria, and viruses [13]. Finally, the activated charcoal filter absorbs gases, odors, and volatile organic compounds.

While effective, this system faces significant drawbacks, particularly with the HEPA filter: Some limitations of HEPA filters include their cost and environmental impact, as they make up 25–35% of the total price of an air purifier. Additionally, HEPA filters are not biodegradable, which means they create waste that often ends up in landfills along with the pollutants they capture [13]. In terms of health and performance, HEPA filters can dry out indoor air, which may worsen respiratory problems, especially in dry climates. Moreover, some purifiers release ozone, a lung irritant, particularly when combined with UV ionization systems. Finally, HEPA filters have difficulty capturing ultrafine particles and certain gases, so they often require extra purification steps to effectively clean the air. To address these issues, mycelium-based filters offer an eco-friendly redesign. Derived from fungal networks, mycelium leverages natural properties for air purification:

- VOC Degradation, wherein, unlike HEPA, mycelium enzymatically breaks down VOCs and toxins, reducing reliance on carbon filters.
- Mycelium is biodegradable and can be grown on agricultural waste, aligning with circular economy principles. This reduces the production costs and waste by 30–50% compared to synthetic HEPA.
- Its porous structure lowers air resistance, enhancing energy efficiency and reducing noise.

5.7 Conclusion

The project explored the human-centric systems design approach to India's mushroom sector, focusing on oyster mushrooms due to their nutritional, environmental, and economic benefits. It detailed the cultivation process, mapping the ecosystem of stakeholders and associated insights into a large gigamap. It also highlighted the industrial applications ranging from food and textiles to bioremediation, exploring vital intervention areas in the mushroom market. The core project intervention focused on mycoremediation—using fungi like oyster mushrooms to break down pollutants. Through secondary research and system mapping, a leverage point was identified; i.e., creating a sustainable air purifier using mycelium as a natural HEPA filter replacement. This biomimetic design improves indoor air quality while promoting circular economy principles, offering a nature-inspired alternative to conventional purifiers. The project demonstrated the steps of how systems design approaches could be successfully used to design a product in the Indian mushroom farming sector.

References

1. Vaidya CS, Sharma ML, Sharma NK (2000) Problems, potential and economics of mushroom cultivation. In: H.P. Agro Economic Research Centre Himachal Pradesh University
2. Mushrooms Production and Harvesting—Penn State Extension. https://extension.psu.edu/forage-and-food-crops/mushrooms/production-and-harvesting
3. Kakraliya SS (2020) Economic importance of mushroom and their uses. Just Agric 1(3):1–8
4. Kulshreshtha S, Mathur N, Bhatnagar P (2014) Mushroom as a product and their role in mycoremediation. AMB Express 4:29. https://doi.org/10.1186/s13568-014-0029-8
5. Journal of Multidisciplinary Science Mikailalsys (2025) Oyster Mushroom in bioremediation: a review of its potential and applications. In: Research Gate. https://www.researchgate.net/publication/387223570_Oyster_Mushroom_in_Bioremediation_A_Review_of_Its_Potential_and_Applications
6. Jain D, Dimri R, Mishra S, Kumar S (2024) Economically important mushrooms of India. https://www.researchgate.net/publication/379443100_Economically_important_mushrooms_of_India
7. Hidalgo J, Epelde L, Anza M, Becerril JM, Garbisu C (2023) Mycoremediation with Agaricus bisporus and Pleurotus ostreatus growth substrates versus phytoremediation with Festuca rubra and Brassica sp. for the recovery of a Pb and γ-HCH contaminated soil. Chemosphere 327:138538. https://doi.org/10.1016/j.chemosphere.2023.138538
8. Johnston MW (2010) Mushrooms offer bioremediation options. In: BioCycle. https://www.biocycle.net/mushrooms-offer-bioremediation-options/
9. Pohan JN, Kusumawati YA, Radhitanti A (2023) Mushroom mycelium-based biodegradable packaging material: a promising sustainable solution for food industry. E3S Web Conf 426:02128. https://doi.org/10.1051/e3sconf/202342602128
10. Wheeler M (2024) MyCoremediation: how mushrooms help clean up the environment. In: R&R Cultivation. https://rrcultivation.com/blogs/mn/mycoremediation-how-mushrooms-help-clean-up-the-environment?srsltid=AfmBOoq-3A9h16tQdP4GwaEP8CF2AC-Y0GK1fN5zVZXpLSOLDgIHksPs

11. Crescent SS, Pisarski EC, Wirth EF, Long RA (2025) Mycoremediation of Louisiana sweet crude oil with Pleurotus ostreatus and nutrient amendments. Environ Toxicol Chem 44:1154–1163. https://doi.org/10.1093/etojnl/vgae078
12. Šrédlová K, Škrob Z, Filipová A, Mašín P, Holecová J, Cajthaml T (2019) Biodegradation of PCBs in contaminated water using spent oyster mushroom substrate and a trickle-bed bioreactor. Water Res 170:115274. https://doi.org/10.1016/j.watres.2019.115274
13. Vaišis V, Chlebnikovas A, Jasevičius R (2023) Numerical study of the flow of pollutants during air purification, taking into account the use of eco-friendly material for the filter—mycelium. Appl Sci 13:1703. https://doi.org/10.3390/app13031703

Chapter 6
Supporting Small-Scale Sellers in Organic Vegetable Markets

Nabajit Bora, Imandi Avinash, and Rahul Mishra

Abstract Vegetable cultivation is a major source of employment as well as a critical aspect of food security. However, the involvement of middlemen significantly reduces the farmers' revenues, often resulting in financial hardship and debt. Despite government initiatives and the advent of online sales platforms, the disconnect between farmers, vegetable retailers, and consumers remains evident. This project aims to aid the farmers in increasing and maintaining their income and improve the experience of buying and selling vegetables by making use of the widely used technology, like WhatsApp™, to create direct farmer-to-consumer links. The methodology for the study entails a thorough examination of the Indian vegetable market system, farmer adoption of technology, and the problems encountered by them. Using ZIP analysis, five essential areas within the vegetable market system were focused on: the current vegetable selling structure, Vegetable waste management, healthy vegetable growth, pollination processes and its effects on vegetable cultivation, and crop damage caused by animals. Enhancing the vegetable-selling structure by bringing sellers and buyers into direct contact was selected as the area for intervention. This approach focused on reducing the farmers' reliance on middlemen, minimizes wastage, and enhances accessibility for farmers and consumers alike. The inclusion of features such as video calling for quality checks would increase trust and ease of use platform, making it a convenient and trustworthy solution for the stakeholders. By leveraging a commonly used digital platform, the proposed solution would help to bring about a change in the supply chain of the Indian vegetable market.

Keywords Vegetable cultivation · Farmers · Middlemen · Supply chain · Online marketplace

N. Bora (✉) · I. Avinash · R. Mishra
Indian Institute of Technology Kanpur, Kanpur, India
e-mail: nabajitb24@iitk.ac.in

V. Kant (ed.), *Human Centred Design for Systems*, Springer Series in Design and Innovation 68, https://doi.org/10.1007/978-3-032-30376-9_6

6.1 Introduction

India is the second-largest vegetable-growing country, that holds 10.6% of the world's vegetable production. The diverse agro-climatic zones enable a year-round cultivation of a wide range of vegetables [1]. A year round production A year-round production generates employment opportunities in various segments of the vegetable sector. On the other hand, vegetables form a staple of any individual's diet in India. The production of vegetables is heavily dependent on multiple factors such as climate, pricing, technology, consumer and market trends, government policies and regulations etc. All these factors are substantially interconnected and together form parts of the vegetable market system. A change in any one of these factors influences the other within the system. For example, a change in access to infrastructure, such as technology in storage facilities, can cause about 30–40% loss in produce between harvest and consumption [2–5].

6.2 Framing the System

Indian agriculture serves as a vital economic pillar and a primary source of employment for a significant part of the population, with the vegetable sector forming an important component of this sector. Vegetables are recognised as essential dietary components that provide necessary nutrients for growth, health maintenance, and overall well-being [6]. In India, where vegetarianism is a core lifestyle, olericulture (the science of growing vegetables) is essential for providing daily nutrition. As vegetables are daily diet staples, their cultivation remains a vital source of both food security and income for small farmers.

Although the challenges in this sector are substantial [7], they cannot be attributed to a single entity. The Indian vegetable market is a primarily social system, rather than a technical one, which relies on limited human contact, or a socio-technical system that integrates humans with technological elements. The traditional Indian market relies extensively on personal relationships, negotiations, and informal understandings that often carry more weight than official policies. In this environment, reputation and bargaining power prevail, with families often engaged in farming, wholesaling, and vending for several generations to maintain their market positions. The influence of government bodies tends to be secondary to these established networks and 'off-the-record' transactions.

For a system as complex as the Indian vegetable market, this holistic systemic perspective allows for an exploration of the root causes and long-term effects rather than only addressing the surface-level symptoms. The initial exploration focused on the vegetable market as a critical subsystem. These markets act as the central bridge between the production and consumption phases of the vegetable life cycle. They often suffer from inefficiencies due to a heavy reliance on middlemen, resulting in lower earnings for farmers and higher prices for buyers [8]. The vegetable market structures vary significantly, where urban centers utilize organized retail and e-commerce, while rural regions rely on direct sales from farmers and small vendors. In urban settings, formal and retail-oriented infrastructures are predominant, often resulting in higher price points due to accumulated transportation, storage, and intermediary costs [8, 9]. Conversely, rural markets tend to have less formal setups and lower prices, where consumers typically perceive the produce to be fresher [10].

6.3 Comprehending the System

The project began with an initial study of the vegetable lifecycle and mapping the stakeholders involved at the different phases of the lifecycle. The stakeholders were categorized as primary, secondary, and tertiary, where primary stakeholders were considered to be those who get in direct contact with the vegetables in their lifecycle, eg. Farmers, Transporters, Wholesales, Retailers, Distributors, and Consumers. Secondary stakeholders were considered to be those who provide support, such as governance, funding, critical services, etc. to the primary stakeholders. They were also considered to be those who have an impact on operations, prices, and infrastructure within the system, such as Service Providers, Agricultural Produce Market Committees, Cold Storage Operators, Financial Institutions, Animals, etc. Tertiary stakeholders were considered to be those who do not directly participate in setting regulations and trade but shape the system through innovation, awareness, and policies, e.g., Government Bodies, Hospitals, Health Professionals, Media Groups, Environmental Groups, Technology Providers, etc.

Visualizing these stakeholders revealed a complex, network-like ecosystem where interconnections define the flow of the produce and its value. While farmers are at the centre of production, they have limited influence over the final prices and profits. Stakeholders like the middlemen, traders, and retailers, in comparison, have a stronger control over distribution and revenue. A careful analysis of these inefficiencies and interdependencies was important to understand the stakeholder roles and the impact each stakeholder would play within the system. Furthermore, shifting the focus from the entire vegetable sector specifically to the Indian vegetable markets helped in identifying certain shifts in the position, role, and influence of the stakeholders. While some remain primary, others shifted from secondary to tertiary, some remained passive, or changed in relevance depending on the market dynamics. For example, in the broader vegetable sector, farmers are the primary producers; however, within the vegetable market system, they also at times function as restricted sellers with limited control over distribution and pricing, remaining heavily reliant on intermediaries. Similarly, while retailers are a part of the consumption phase in the wider system, they act as immediate demand drivers within the vegetable market, influencing consumer trends and sourcing methods. Finally, while consumers may appear as passive end-users with indirect influence on the prices in the wider vegetable sector, they directly influence market price and availability through their demand for convenience, organic options, and affordability, forcing wholesalers and retailers to respond accordingly in the context of the vegetable market.

Next, a stakeholder power analysis was done to identify the interdependencies and inefficiencies between different stakeholders. This analysis helped understand the role and impact of these stakeholders within the system, for example, farmers (primary stakeholder), the 'producers' of vegetables have limited control over the final price and distribution of the produce as they are reliant on the middlemen and are thus more vulnerable to market fluctuations. The middlemen (secondary stakeholder) are gatekeepers between the farmers and retailers and thus have significant influence over the final price of the produce. The retailers are the final link between the middlemen and consumers and have the most control over the final price of the produce. In contrast, the consumers' control over the prices is passive and indirect, where they are not able to set the final price of the produce; however, their purchasing behaviour and decisions affect the market trend, demand, and, in turn, affect the prices. By tracing these roles and power shifts, the stakeholders who have the differing potentials to affect the system were identified.

To understand the variables and constraints under which the stakeholders were operating within the system, the stakeholder dimensions were mapped. Taking the example of a farmer, the first variable that was mapped was 'land size', where it could span from a minimum of 0.5 acres to a maximum of 100+ acres. This indicated whether the farmer was operating on a small/medium/large scale. Small-scale farmers use APMCs (Agricultural Produce Market Committee), while large-scale farmers can bargain for better prices with the middlemen or sell directly to the retailers. Another variable that was mapped was resources, specifically funds. The amount of funds available would have a direct impact on several aspects of farming, right from the selection of crops to cultivate, equipment to be used, manpower to be hired etc. Other variables that were mapped out were water requirement, accessibility to water resources, electricity requirements, seed costs, labour costs, access to markets, etc.

The mapping of dimensions extended to other primary stakeholders, such as sellers (wholesalers, commission agents, and traders), whose variables included warehouse space, daily sales volume, and pricing control. Distributors were analysed through variables like farmer network, distance covered, spoilage loss percentages, fuel costs, and cold storage dependency, while consumers were defined by income levels, purchase frequency, consumption behaviour, preference for organic vegetables, bargaining frequency, etc. Finally, the dimensions covered for cold storage providers were storage capacity, energy consumption, storage rent, temperature range, spoilage rates, etc. The variables for key secondary stakeholders, including seed and pesticide manufacturers, financial institutions, and waste management companies, were also detailed. These variables showed significant interdependency; for example, the extent of a farmer's holding would determine their ability to withhold stock for better prices, while a wholesaler's network size would dictate their influence over setting prices. Plotting these dimensions across the system highlighted exactly where the greatest inefficiencies and inequalities existed, providing a foundation for more effective interventions.

Further, the Indian vegetable market's hierarchical structure was mapped. The Indian vegetable market relies extensively on inter-personal relationships, negotiations, and informal understandings. The hierarchy was divided into five levels to define the social structure of the vegetable markets. At the 'Individual Level (lowermost level)' were farmers, wholesalers, retailers, and consumers, as each can be independently dependent on factors like personal earnings, bargaining power, resource availability and accessibility, and local market forces. For example, a street vendor can choose what vegetables to trade on the basis of perishability and profitability. The household economic structure was considered at the 'Family Level' where the majority of farmers, merchants, and shopkeepers have family businesses that have generational knowledge, trade techniques passed down, and also affect the scalability of the businesses. At the 'Community Level', farmer organisations, wholesale trader networks, and mandi communities were considered. At the 'Institutional Level', APMCs, Banking Networks, and transportation infrastructure were considered. At the topmost level of the social structure - 'Society Level', food consumption patterns, economic changes, urbanization, environmental factors, government policies, global trade, etc., were mapped. The hierarchical breakdown helped envision how a change at any level may impact the rest of the levels. For example, a farmer performing poorly at the individual level will have an effect on family stability, market supply chains, and eventually even affect government policies.

Moving ahead, an exercise mapping the primary actors and their roles in the system was conducted. The actors in the system interact and affect each other in multiple ways. The influence of these actors, along with other factors such as social, economic, and cultural practices, can impact the system both positively and negatively (Table 6.1). The most apparent players observed were the farmers, traders, transporters, and consumers. Categorising these influences as positive and negative brought about an understanding of the power dynamics. For example, the middlemen and wholesalers have a higher degree of power to influence the economy, whereas the farmers and consumers have the least influence. While these players are the most visible, their influence and roles don't tend to be equal. Charting these differences revealed that the middlemen possess the most economic power, as they set prices and manage supply chains. As opposed to that, the farmers occupy the lowest place on the economic ladder. They earn the least despite doing the most labour. Furthermore, while supermarkets and retailers are gaining increasing power in cities, they still remain heavily dependent on wholesalers for their inventory.

Consumer behaviour also plays a key role in shaping the market. Urban consumers often favour packaged vegetables, encouraging the dominance of supermarkets, whereas rural consumers continue to purchase from local vendors and mandis, thereby contributing to preserving traditional supply chains. The resistance to fully change is also seen among small-scale farmers, who often rely on generations of experience and tend to avoid new AgriTech innovations. Additionally, cultural and religious practices introduce unique constraints, such as the avoidance of certain vegetables like onions in some households, which directly impacts local demand. This mapping also distinguished between the visible and hidden effects of the actors. Economic structures, institutional frameworks, and long-term trends are

Table 6.1 Key actors and their positive and negative effects

Actor type	Positive effects	Negative effects
Institutional	Ensures regulation & market control	Bureaucracy slows reforms
Long-term trends	Drives efficiency & innovation	Small vendors struggle to adapt
Economic	Supports supply chain investment	Price manipulation benefits middlemen
Practice	Allows market flexibility	Farmers face financial insecurity
Culture	Preserves farming traditions	Resistance to change slows

visible from the outside. However, the hidden effects such as ingrained practices and cultural norms, are less apparent but can vary across different regions. Mapping these effects demonstrated that while economic influence is concentrated in the hands of middlemen and wholesalers, social and cultural norms also play a major role in influencing the system.

Having identified the key stakeholders, hierarchies, actors, and their influences, a process map for the vegetable market was developed. The process was broken down into four basic, important, simpler aspects: Seed, Farming, Vegetables, and Consumption. 'Seeds' were considered the first aspect of the vegetable market. The selection of an appropriate seed type determines the resistance, protection against pests, and harsh weather a crop would have. It also influences the nutritional value, shelf life, and taste of the vegetables. GMO (Genetically Modified Organism) seeds altered for high resistance and more yield represent a fundamental shift in agricultural practices. These seeds create loss of biodiversity, market monopolies, and ultimately redirect the farmers' income towards multinational corporations developing these seeds. Hybrid seeds, though, are more productive than the regular seeds; they cause the farmers to spend more as they need to be bought fresh every season. Heirloom seeds, which are known for their original taste, have the drawback of producing a lower yield. Lastly, higher upfront costs for organic seeds (free from synthetic fertilisers and pesticides) deter farmers despite their environmental and sustainability benefits. Most of the small-scale farmers who are unable to afford good-quality seeds depend on government subsidies or traditional preservation methods. Community seed banks prove to be an alternative solution to this, where farmers borrow seeds at the start of the season, and return a larger quantity of seeds as repayment. This circular exchange model provides aid to the farmers while ensuring biodiversity. In situations where the seeds are sold instead of being given on loan, seed adulteration is observed. Poor-quality seeds are sold at inflated prices, resulting in substandard crop harvest and financial hardship for farmers.

'Farming' was considered the second aspect of the process map. This aspect covers activities done after seed selection till harvest, i.e., sowing, adding fertilisers and manure, irrigation, weeding and protection, and harvesting. Conventional farming methods utilise chemical fertilisers and pesticides that produce better yields; however, they also cause long-term soil deterioration. On the contrary, organic farming promotes the use of natural pest control and composting, which are more environmentally sustainable but produce lower yields. Permaculture and natural farming methods replicate nature, minimising external support. While this method creates self-sustaining ecosystems, improving the environment, farmers tend to find it challenging to adopt due to its steep learning curve. 'Vegetables' were considered the third aspect of the map. This aspect focused on collection, distribution, and supply chain activities related to vegetables. One of the important factors that influences the collection and distribution activities is the storage of vegetables. The vegetables are sorted and graded based on size, colour, and quality. Root vegetables, such as potatoes, carrots, and onions, are less perishable and can be stored for months. Leafy vegetables such as spinach, coriander, and fenugreek are highly perishable and tend to require refrigeration. Fruiting vegetables such as tomatoes, chillies, and brinjals require rigid storage that prevents them from getting crushed during transportation. Absence of proper storage infrastructure results in significant losses.

The last aspect considered for the process map was 'Consumption'. This aspect can be categorised into two types: Household Consumption and Processed Food Industry Consumption. In households, vegetables are consumed daily, and hence, consumers prefer buying them fresh on a daily or weekly basis. In the food processing industry, vegetables are purchased in bulk and used in frozen meals, pickles, sauces and snacks. Depending on the kind of product that is to be produced, vegetables are purchased on a contract basis either seasonally or off-season. The process map highlighted the interdependencies within the system, where small inefficiencies can cascade into larger issues impacting everything from farmers' revenue to overall food security.

To understand the difficulties that may arise in the system, challenges across the different stages of the vegetable lifecycle were mapped. Each stage would have its set of intricacies, inefficiencies, and systemic deficiencies that impact the overall system. Price fluctuation is a critical challenge that has a significant influence on the system, and is also influenced by the system. Price is affected by multiple factors, such as seasonal fluctuation, weather unpredictability, export-import dynamics, market speculation, etc. Seasonal fluctuation causes prices to drop when harvests are huge and to spike when the off-season creates scarcity. Unexpected changes in the weather, such as rainfall, droughts, and floods, alter the produce supply, consequently affecting the prices. Export bans reduce prices, saving consumers' money, but they also lower farmers' profits, while import duties keep prices high to protect local farmers from foreign competition. Market speculation by middlemen and wholesalers hikes prices when they stockpile vegetables to create artificial shortages. In mandis (wholesale agricultural markets), prices are determined by middlemen who charge commissions, often eroding farmers' profit margins. Farmers tend to receive about 20–30% of the retail price, while the middlemen get a larger percentage. In India, the multi-layered agricultural supply chain, from APMCs (Agricultural Produce Market Committee) to commission agents and retailers, adds high margins that significantly inflate consumer prices compared to what farmers receive [8]. In addition to price fluctuation, challenges related to waste management, post-harvest losses [11], poor storage facilities, poor handling of vegetables, and pollination constraints [12] were

Fig. 6.1 Final gigamap with all the interconnections

also identified. A final gigamap established the links between the different maps and highlighted the parts within the system that are interconnected (Fig. 6.1).

6.4 Envisioning the Interventions

To identify crucial areas of intervention within the Indian vegetable market system, the process ZIP analysis was followed. 18 Zoom-in points were initially identified, which were further narrowed down to 5 Zoom-in points. They mentioned as follows–

- Zoom-in Point 1–'Indian Vegetable Market Selling Structure' was selected as the selling structure is highly reliant on middleman, with the farmers getting a significantly smaller share of the end price [13].
- Zoom-in Point 2–'Post-Harvest Processing and Waste Management' was selected as a huge part of the vegetables in India are wasted due to spoilage and ineffective logistics [11].
- Zoom-in Point 3–'Organic Farming & Health-Priortised Vegetables' was selected because there is low organic farming adoption despite consumers becoming health-conscious [14].
- Zoom-in Point 4–'Increasing the Pollination Process' as disruptions to the pollination process directly reduce vegetable yields [12].
- Zoom-in Point 5–'Animal Destruction of Crops' was selected as farmers suffer significant losses when crops are destroyed by wild animals [15].

Based on the 5 Zoom-in points, 5 Intervention points were proposed respectively. They are mentioned as follows -

Intervention Point 1: Vegetable markets' selling structure and experience.
Intervention Point 2: Waste treatment & usage.
Intervention Point 3: Awareness of health-prioritised vegetables.
Intervention Point 4: Natural pollination process and techniques.
Intervention Point 5: Safeguarding of crops from animals.

Having determined the five key intervention points, the most effective leverage point was identified. A cost-to-time impact analysis was conducted, plotting the feasibility of each solution against its potential effect. To identify high-impact interventions requiring the fewest resources, the intervention points were plotted on the matrix using time for implementation (X-axis) and cost (Y-axis). Each intervention was ranked based on an approximate estimation to envision its implementation. 'Restructuring the vegetable market structure' was identified as the most impactful leverage point, based on which 5 possible solutions were conceptualised. They are mentioned as follows–

1. A WhatsApp™–based Marketplace
2. Card tagging for vegetables with details
3. Specialised app for tracking vegetable carts and their live location
4. Live tracking of vegetable pricing and availability status
5. Mud and solar-powered storage units

To narrow down and select one solution out of the five, a cost-to-time graph was plotted again. A WhatsApp-based Marketplace was selected as the final solution as it showed the potential to have a greater impact on the system and would require less time to implement.

6.5 Formulating the Interventions

The proposed solution, 'TazaBazar' (Fresh Market) would be a WhatsApp™ - based plugin that would connect farmers, retailers, and consumers directly through an accessible interface, ensuring a seamless, reliable, flexible, and user-friendly experience [16].

WhatsApp™ is a commonly used mode of communication by a majority of Indian farmers. Leveraging familiar platforms such as WhatsApp™ for agricultural marketing reduces adoption barriers [17]. Farmers would register through WhatsApp by installing the TazaBazar plugin. They would upload details such as type, quantity, price, as well as the pictures of the vegetables for sale. The retailers and consumers would receive real-time alerts about fresh stocks available in their area, which could be filtered by the 'nearby radius' feature. Direct orders would be placed via video call, and delivery would be facilitated either by delivery partners or the consumers would have the choice to opt for 'self-pickup' or 'direct delivery by sellers'. Lastly, a flexible payment system would be provided with 'cash-on-delivery payments' or 'pay-later' options, ensuring seamless transactions between farmers and consumers.

Four personas, each representing different stakeholders who would potentially make use of the proposed solution, were developed. They are mentioned as follows -.

Persona 1: Ramesh, a 42-year-old small-scale vegetable farmer from Uttar Pradesh, farms seasonal vegetables on 2 acres of land. He feels that, despite his hard work, the middlemen take most of his profit. He wishes to increase his profit margin by selling directly to the consumers.

Persona 2: Sunita, a 38-year-old Kanpur-based retail vegetable vendor, sells vegetables in her small roadside shop. She feels that the prices of vegetables change everyday and affect her business. She often ends up with unsold stock and wishes to manage her stock efficiently.

Persona 3: Aravind, a 29-year-old high school teacher from Kanpur, is punctual about his work and conscious about his health. He finds it difficult to trust and buy fresh and organic vegetables regularly.

Persona 4: Ahmed, a 29-year-old delivery partner from Uttar Pradesh, delivers goods locally. He finds long distances, bad roads, low pay and uncertain income challenging. He wishes to regulate his delivery routes and manage his income accordingly.

These personas provided insight into how different users would interact with the proposed solution and navigate to their goals, as well as the potential gaps and difficulties they might encounter. The information architecture was built to work within the existing WhatsApp interaction constraints. The top-level of the information architecture included the initial entry point for the user, i.e. the user registration, wherein the farmers, retailers, and consumers would select the user type, and give necessary permissions to move forward through the process (Fig. 6.2). The onboarding would take minimal time as native preferences such as location, contact details, and language from WhatsApp would be taken. Based on the final information architecture and the user personas, the user flows for sellers, delivery partners, and consumers were developed (Fig. 6.3).

Considering the core functionalities and navigation, the initial steps for the farmers would be to add the vegetable details (picture, quantity, price), see the order and delivery requests, confirm direct orders via video call, and lastly look at everyday demand trends and suggested prices. On the other hand, for retailers and consumers, the initial steps would include viewing the vegetable status updates in real time, searching and filtering based on locality, price, and freshness, then placing orders, and lastly scheduling deliveries. For delivery partners, the initial step would be to accept and schedule pick-ups from farmers, track the delivery location and optimize delivery routes. The UI design of TazaBazaar was designed to focus on simplicity, accessibility, and clarity to ensure a seamless experience for both farmers and consumers. This user-centric approach guided the process from initial low-fidelity sketches to final high-fidelity mockups. Based on initial user feedback, the interface was refined to reduce steps and eliminate friction, ensuring the platform remains intuitive even for those users who might have limited technical proficiency (Fig. 6.4).

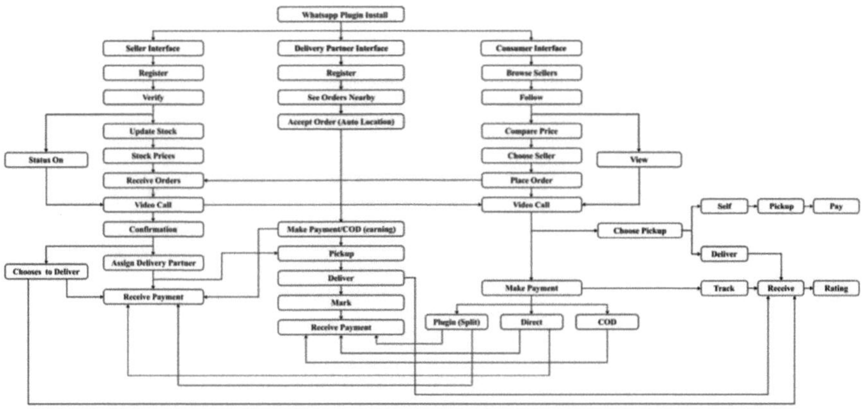

Fig. 6.2 Final information architecture of the plugin

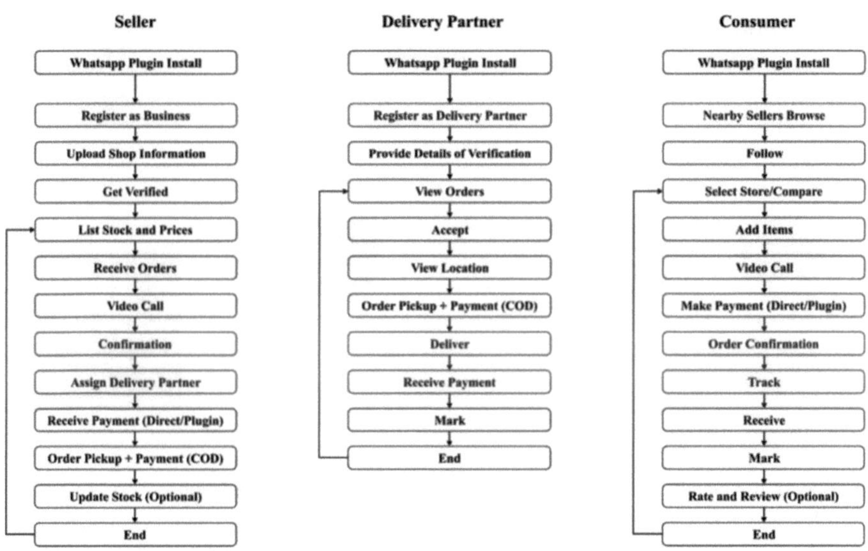

Fig. 6.3 User flows for all the 3 prominent stakeholders (sellers, delivery partners, consumers)

The branding for Tazabazar was designed based on two fundamental aspects. The first one considered was freshness and trust. Freshness was reflected in the brand name-'Tazabazar', where 'taza' means 'fresh', and trust was reinforced by using the tagline 'Connecting farmers to consumers, one click at a time'. The second aspect was to connect with the audience through a direct, friendly image. This was reflected in the logo, which depicts a cart with wheels to symbolise the concept of reaching consumers directly at any time. Green was incorporated in the color scheme that reflected sustainability, freshness, trust and reliability (Fig. 6.5).

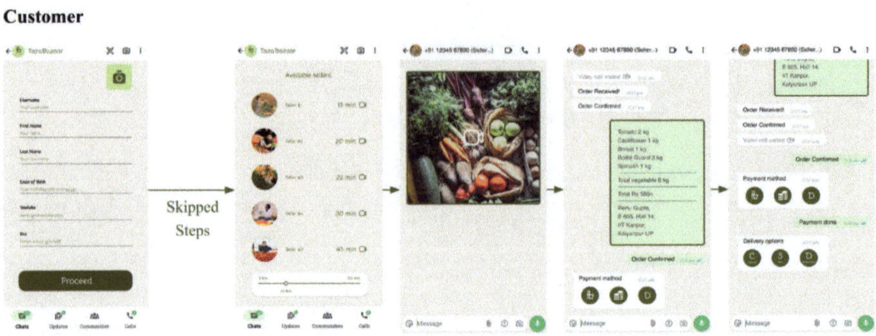

Fig. 6.4 Final screens for the consumers and the process steps involved

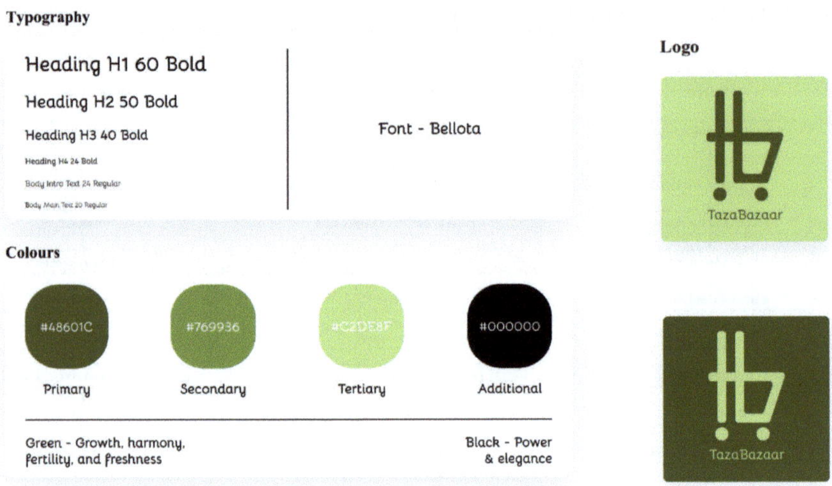

Fig. 6.5 Brand identity for the plugin TazaBazaar along with initial iterations

The proposed solution–Tazabazar, encompasses a direct farmer-to-consumer (F2C) model with 3 specific roles defined modes (Fig. 6.6). It addresses the following challenges:

- Minimizing vegetable market inefficiencies by reducing dependency on middlemen, resulting in greater farmer profits.
- Providing consumers with fresh, affordable vegetables, thereby curbing price inflation.
- Optimizing supply chain efficiency to significantly reduce food wastage.
- Establishing a transparent pricing mechanism that empowers small-scale farmers.

Fig. 6.6 Three key mode options for the plugin based on stakeholder role

6.6 Enabling the Interventions

The adoption process for the proposed solution would begin with a pilot program to gather feedback and make iterative UI/UX improvements before a wider rollout. To build trust and reliability, collaborative government campaigns for onboarding would be launched. Offline awareness in rural areas would be driven by publicity at key centers like mandis and community meetups to reach the last-mile users. The monetization plan would include a minimal platform fee and small commissions charged through a plugin. To encourage growth, a referral reward system would be introduced to incentivize farmers to invite others. At a later stage, in-app advertisements could even provide farmers with awareness of various government initiatives. The release strategy would start with a Minimum Viable Product (MVP) and rigorous testing. To sustain long-term trust, a word-of-mouth strategy will capitalize on early adopters and farmer cooperatives. This would be followed by expansion through partner-led promotions and local meetings. Finally, TazaBazaar would scale across various locations in India, with a future chatbot implementation to further simplify the user journey.

6.7 Conclusion

The final synthesis map (Fig. 6.7) illustrates a comprehensive process by integrating stakeholder, actor, and hierarchy mapping with task, process, and challenge analyses, moving from left to right, narrowing down on areas that require intervention and deriving a final systemic solution. This project envisions a restructured agricultural landscape where farmers could bypass traditional intermediaries to sell their produce directly.

By leveraging commonly used digital platforms like WhatsApp™, farmers would be enabled to list fresh vegetables, set competitive prices, and connect instantly with

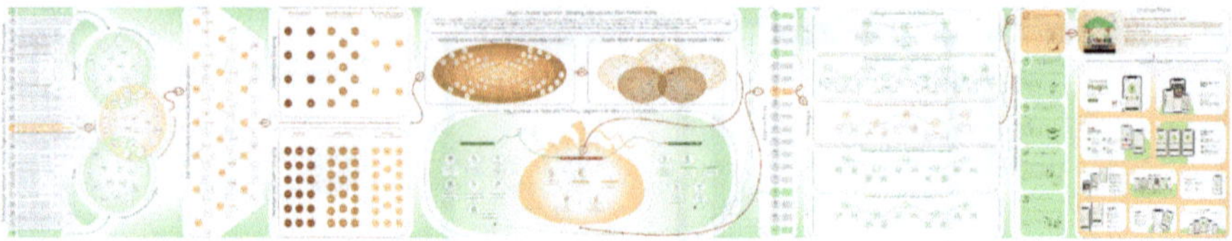

Fig. 6.7 Final synthesis map depicting overall process

local consumers. The vegetable supply chain would become increasingly resilient and self-sustaining as more farmers join the ecosystem. This initiative represents a systemic shift toward fair trade and improved livelihoods rather than a simple digital application. Through a systems-oriented approach, the project bridges the gap between producers and consumers, creating a scalable impact that redefines the Indian vegetable market as a more efficient, farmer-centric environment.

References

1. Research K (2025) Major trends transforming India's vegetable market in 2024. In: Ken research. Ken Research. https://www.kenresearch.com/blog/emerging-trends-in-india-vegetable-market. Last accessed 20 Mar 2026
2. Ranjan J, Sahni R (2023) Post harvest losses of fruits and vegetables in India. Indian Agric
3. Kohli P (2017) Analysing NDDB cluster model for marketing of vegetables. In: Cold chains, vol 1. National Center for Cold-chain Development, New Delhi. https://nccd.gov.in/PDF/Analysis_NDDB_veg_model.pdf
4. Kumar S, Meena P, Kumar R, Venkateshwarlu G (2023) Sustainability of vegetable street vendors in India. In: National Academy of agriculture research management. ICAR-NAARM, Hyderabad. https://naarm.org.in/
5. Sachdeva S, Sachdev TR, Sachdeva R (2013) Increasing fruit and vegetable consumption: challenges and opportunities. Indian J Community Med 38(4):192–197
6. Kishore GR, Singh N (2017) Role of vegetable (Olericulture) production in Indian economy. In: Advances in plant sciences. Academy of Plant Sciences, Allahabad
7. Merriott D (2016) Factors associated with the farmer suicide crisis in India. J Epidemiol Glob Health 6(4):217
8. Dastagiri MB (2013) Indian vegetables: production trends, marketing efficiency and export competitiveness. Am J Agric For 1(1):1
9. Hegde RN, Madhuri NV (2013) National Institute of rural development a study on marketing infrastructure for fruits and vegetables in India. In: Research reports series. National Institute of Rural Development Hyderabad, p 91
10. Ideas for India (n.d.) Exploring urban perceptions around freshness of vegetables. https://www.ideasforindia.in/topics/agriculture/exploring-urban-perceptions-around-freshness-of-vegetables. Last accessed 20 Mar 2026
11. Kaur R, Kaur B (2024) Post harvest management of vegetables: a review. Agric Rev 45(3):520–525
12. Basu P, Bhattacharya R, Ianetta P (2011) A decline in pollinator dependent vegetable crop productivity in India indicates pollination limitation and consequent agro-economic crises. Nat Prec
13. Thangamani (2016) Indian agriculture: performance and challenges—a case study. Shanlax Int J Arts Sci HumIties

14. IMARC Group (2024) India organic food market: industry trends, share, size, growth, opportunity and forecast 2025–2033. IMARC Group, Noida
15. Hazra P (2023) Trends of crop damage by wild animals in Darjeeling Hills, West Bengal, India. Asian Bus Rev 13(2):84–90
16. Thakur D, Chander M, Sinha S (2017) WhatsApp for farmers: enhancing the scope and coverage of traditional agricultural extension. Int J Sci Environ Technol 6(4):2190–2201
17. Balaji G, Kavaskar M (2024) An analysis of farmer's perception towards WhatsApp as a farm technology dissemination tool. J Exp Agric Int 46(9):330–335
18. ICAR (2024) Indian Council of Agriculture Research-Annual report 2023–24. https://icar.org.in/annual-report. Last accessed 14 Feb 2025
19. Meadows D H M (2004) Dancing With Systems. Timeline: A Publication of the Foundation for Global Community-Current Issue, Last accessed 14 Jan 2025
20. Sengupta R (2024) Agriculture subsidies in India. International Institute for Sustainable Development, Winnipeg. Last accessed 10 Feb 2025
21. Kumar R (2023) Farmers' use of the mobile phone for accessing agricultural information in Haryana: an analytical study. Open Inf Sci 7(1)
22. Ramaswami B (2019) Agricultural subsidies. Study Prep XV Financ Comm New Delhi: Indian Stat Inst. https://fincomindia.nic.in/archive/writereaddata/html_en_files/fincom15/StudyReports/Agricultural%20subsidies.pdf
23. Kappa K, Anuradha C, Praveen Kumar B (2022) Marketing cost, channels and constraints: a case study of vegetables marketing in Telangana State. J Res ANGRAU 50(2):134–146
24. Singh AK, Yadav A, Kumar V, Halim MA (2023) Market access for organic and natural produce case studies. In: Shankar A (ed) Centre for science and environment, New Delhi. https://www.cseindia.org/market-access-for-organic-and-natural-produce-case-studies-11734

Chapter 7
Enabling Endurance of Athletes

Ayantika Roy, Muskan Agrawal, and Vivek Pal

Abstract The future of endurance in Indian athletes is a crucial and under-researched area that directly impacts sports performance, career longevity, and overall well-being of athletes. As India continues to make strides in global sporting events such as the Olympics, Commonwealth Games, and professional leagues, the need for scientifically backed endurance training and recovery strategies has become more evident. However, despite the growing participation in competitive sports, Indian athletes often face significant challenges, including inadequate recovery protocols, risks of overtraining, and limited access to advanced sports science methodologies. We use a design approach in terms of integration of technical textiles and soft sensor technology to transform athletic recovery and injury prevention in India. Unlike conventional training gear, this design concept combines real-time muscle health tracking, adaptive temperature regulation, and proactive injury prevention into a single wearable system. The incorporation of conductive fibers and shape memory alloys ensures optimal body temperature management, reducing muscle fatigue and enhancing endurance during intense training sessions. By embedding soft sensors within the fabric, the product continuously monitors biomechanical and physiological data, enabling athletes and coaches to make informed, data-driven decisions. This design concept not only accelerates recovery and minimizes injury risks but also revolutionizes training methodologies by providing personalized insights tailored to each athlete's needs. With India's growing presence in global sports, such advancements are crucial for improving athletic performance, extending career longevity, and strengthening the country's sports ecosystem. By making cutting-edge recovery solutions accessible to both elite and grassroots athletes.

Keywords Endurance · Athletes · Gigamapping

A. Roy (✉) · M. Agrawal · V. Pal
Indian Institute of Technology Kanpur, Kanpur, India
e-mail: ayantika24@iitk.ac.in

V. Kant (ed.), *Human Centred Design for Systems*, Springer Series in Design and Innovation 68, https://doi.org/10.1007/978-3-032-30376-9_7

7.1 Introduction

Running is a popular and accessible sport worldwide. It's fundamental for many other sports, catering to all athletic levels, and enhancing general fitness. It's an ideal focus for improving endurance and impacting a broad range of people who are not just athletes but also try to maintain their physical health, ranging from beginners to advanced levels. Endurance is one of the most critical factors for athletic performance, especially for runners. Endurance is the ability to sustain physical activity or even mental effort over an extended period. For athletes, this could mean running longer distances without losing speed or maintaining focus during intense training sessions. Endurance can be seen as a combination of stamina, ability to keep going, and resilience, having the ability to recover from fatigue or discomfort. For example, a marathon runner who maintains a steady pace for 42 km demonstrates exceptional physical endurance.

7.2 Framing the System

To comprehend and design for endurance, firstly the need to explore what endurance is and why it matters for athletes in India, and how a human-centered systems design approach can systematically improve it. So, endurance is not just about running longer; it's about running smarter. It allows athletes to perform better over extended periods, recover faster between training sessions, and build mental toughness for competition. For instance, an athlete with high endurance can complete a 10 km race without slowing down in the final kilometers, which is a critical advantage in competitive sports. In India's evolving sports landscape, focusing on endurance can unlock untapped potential among athletes who often face challenges like limited resources or recognition. For example, take a look at Milkha Singh; despite his hardships, he excelled in the 200-m and 400-m races at the 1958 Asian Games and Commonwealth Games. His performance at the 1960 Rome Olympics showcased his ability to maintain speed over extended periods, inspiring generations of Indian athletes to focus on endurance. Understanding who influences an athlete's endurance journey is important for creating effective strategies that support their development.

7.3 Comprehending the System

To begin with, identifying the main factors which affect the endurance of athletes, being Training [1], Diet and Nutrition, Healthcare [2], Government Policies, and Environment [3]. Based on these, classifying different stakeholders as Primary, Secondary, and Tertiary on their level of impact and interaction with the Athletes as seen in Fig. 7.1. The primary stakeholders are the ones who directly impact an

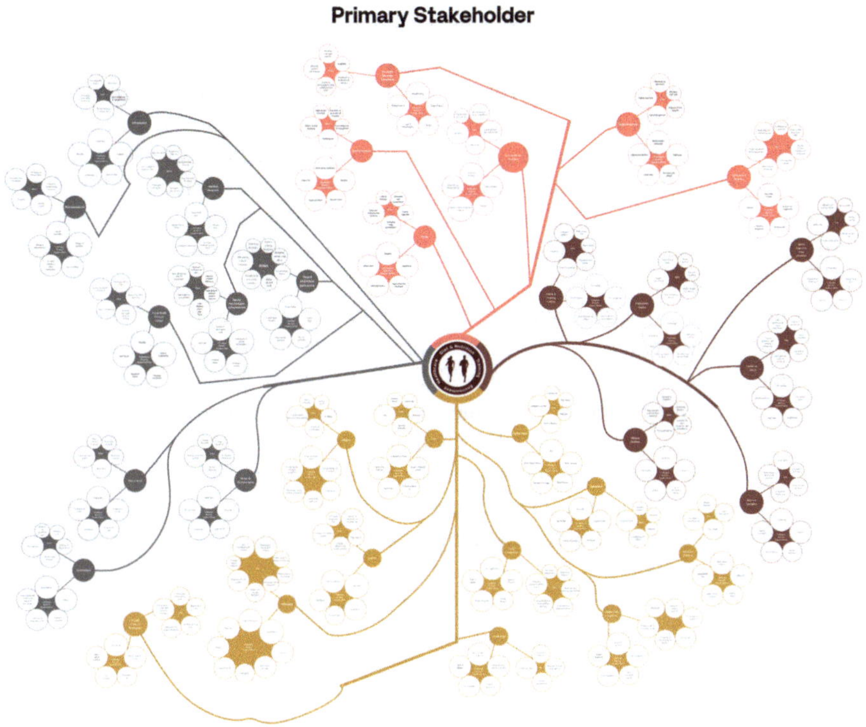

Fig. 7.1 Stakeholder mapping involving primary, secondary and tertiary stakeholders

athlete's performance, secondary stakeholders are the ones who provide necessary support, and tertiary stakeholders are the ones who shape the broader environment and base within which these athletes operate.

Executing this step is important to understand what roles they play in an athlete's life, as well as what challenges and barriers an athlete faces, which, even after the stakeholder's involvement, many times are often not solved and affect an athlete's performance. So by recognizing these challenges within the stakeholder mapping framework, highlighting the main pain points and the gaps, which later on helped in developing targeted strategies to enhance support systems around the athletes.

Moving forward, thc ncxt stcp was to map out all the factors that are directly or indirectly affecting endurance for running athletes, as they often go unnoticed. Mapping these systemic influences, which are interdependent and interconnected with endurance, involved five steps. Delving into the systemic influencers to understand their scope and impact by identifying ecosystem actors that interact and contribute to the overall functioning, including individuals, organizations, or entities that play a role in shaping Endurance. Plotting stakeholders based on their level of influence and proximity to athletes gives an overall understanding. Further analysis of the emergent factors shows patterns or issues arising from systemic interactions, and grouping the main factors to create actionable insights.

Firstly, Systemic Influencers for running athletes, which include Institutions and Governing Bodies that set policies; Training programs that build stamina; Diet and Nutrition strategies for fueling performance; Healthcare systems for injury prevention and recovery; and Environmental factors such as air quality or climate conditions. The next step is to thoroughly analyze each systemic influencer to understand how it impacts endurance. This involves identifying what each factor affects, how it interacts with other factors, and why it influences. By diving deeper into these relationships, it's easy to uncover the mechanisms and interdependencies that drive endurance outcomes.

This mapping revealed some important insights. Some of the phenomena observed can be labelled as "emergent". Emergent phenomena are patterns or issues arising from interactions within the system. These are critical insights that arise from analyzing stakeholders and other influences. These are not the results of the parts themselves but emerge as a holistic phenomenon. The emergent factors include certain gaps that, even after the stakeholders involvement, were not completely solved. Psychological stress, Gastrointestinal issues, Financial stress, and a few others. All these were the hidden challenges that are not immediately apparent to the outside world in general.

Recognizing these emergent factors helps in understanding where they can actually be zoomed into, so that they can detail out each aspect of these issues, which would include their causes, the existing methods that an athlete follows, and details about their environment, so that they can look at the bigger picture. Zooming into the scale of daily life, to understand what it means to be an Athlete. It gives a rich, qualitative understanding of the athlete's routine, right from waking up in the morning to going back to sleep at night. But why was it important to study the entire day of an athlete? As performance isn't just about what happens during training, it's about every single choice, every single moment. This isn't about generic fitness advice; it's about understanding why athletes do what they do and how that all adds up to creating a champion. To analyze the daily life habits and routines of different athletes, as well as the activities that occur within their institutions. These include wake-up schedules, pre-training routines, Training Routines, post-training periods, Meal Schedules, Yoga Practices, high-intensity workouts, Recovery Periods, and the necessary recreational habits. All these elements work together to ensure sustainable performance over time while promoting physical health and mental well-being [2].

Measuring endurance isn't as straightforward as it seems; it varies across sports and different individuals, for which several metrics are analyzed. For instance, differences occur due to gender, altitude, climatic conditions, or even nutrients. One can also understand by an example, like how a swimmer's endurance metrics differ significantly from a runner's. Commonly used measures include VO2 Max, which is the maximum amount of oxygen one's body can use during exercise, and lactate threshold, which is the point at which the muscles begin to fatigue due to lactic acid buildup. However, there's no universal standard that is specifically tailored for running athletes that too especially in the Indian context, making it hard to benchmark progress. As part of the mapping process, a variable chart had for the extremities that can support and provide information to the athletes about different levels of body nutrient levels, as well as training precaution levels for proper guiding and to avoid future mishaps. The variables map provides data on different dimensions, such as considering weights, height, intensity, intake of different nutrients, and many others, for benchmarking.

Further, focusing on equipment-based training because training in general not only involves the physical aspect but also includes focusing on gyms and fitness centres, which affect muscle strengthening and endurance using equipment, which are the key factors for long-distance running. It also reduces injury risk by strengthening stabilizer muscles like the glutes and core, which are often neglected in traditional running drills. To tackle this, the approach involved three key steps. First, identifying commonly used equipment such as treadmills, stair-steppers, and resistance bands. Next, moving on to studying biomechanics to understand which muscles are engaged during their use. For example, treadmills primarily target lower body muscles such as calves and the knee joints. Lastly, these activation patterns were mapped as detailed charts to visualize their impact on running performance and understand which are the primary muscles to focus on for training, which came to be the Glutes, Hamstrings, and Quadriceps. These findings are crucial as they can help runners choose equipment that aligns with their training goals accordingly. From training, shifting the focus to explore the often-overlooked yet essential components of training, which are warm-up, cool-down, and rest. These pre-training and post-training methods are essential for maximizing performance, preventing injuries, and aiding recovery. However, they are frequently neglected by athletes to explore why these practices are so important, the physiological benefits of warm-up and cool-down routines, different methods of executing them, and the consequences of skipping them. A good warm-up is like starting a car engine on a cold morning, as it prepares the body for action. It increases blood flow to muscles, ensuring they receive enough oxygen for optimal performance. It also raises heart rate gradually, reducing stress on the cardiovascular system during intense exercise. Cooling down is just as critical as warming up because it prevents sudden drops in heart rate that can lead to dizziness or fainting. It also helps remove lactic acid buildup in muscles, which reduces soreness and stiffness after exercise. Finally, cooling down promotes relaxation by gradually lowering the body temperature.

What happens if one neglects these things? Neglecting these practices can have serious consequences, such as skipping a warm-up increases the risk of injuries like muscle strains because cold muscles are less flexible and more prone to tearing under stress. Similarly, skipping a cool-down can lead to poor recovery, causing fatigue or even overtraining syndrome over time [4]. Through the training analysis, many injuries stem from bad posture or incorrect running form, can be visualized. These issues not only lead to injuries but also slow down improvements in endurance. To understand further, starting with studying correct posture techniques, the approach chosen involved five steps. Firstly, identification of the key elements of correct running posture, which include maintaining an upright torso, keeping shoulders relaxed, and maintaining a slight forward lean from the ankles. Next, analyzing factors that affect running ergonomics, which include Muscle imbalances, Improper footwear support, Poor flexibility in muscle areas, and Environmental factors such as uneven terrain or extreme weather. Then further mapping posture to specific body muscles. For example, maintaining an upright torso engages the core and lower back, thereby keeping the shoulders relaxed, which further activates the upper back and shoulders, and a slight forward lean engages the core, glutes, and calves. Corrective techniques were implemented, which include certain strengthening and flexibility exercises, proper footwear selection, and focusing on posture and ground contact. Finally, assessing the importance of these techniques for improving running performance by preventing injuries and improving efficiency.

Next, a ZIP analysis was conducted with a focus on healthcare, where professionals with specialized knowledge in sports medicine and performance nutrition play a vital role. ZIP is an acronym for Zooming in, identifying pain points, and then formulating interventions. In this case, the ZIP analysis provided insights into certain important concerns. While endurance training has many benefits, it can be harmful if performed at high intensity or volume. The expertise of medical professionals is essential for optimizing an athlete's overall performance, physical health, and mental health. To understand the healthcare system, the key stakeholders were identified and categorized accordingly. Primary stakeholders, such as specialized Doctors related to Training, Diet and Nutrition, Hospitals, Physiotherapists, and Recovery Specialists, play a vital role in injury prevention and rehabilitation, while Secondary stakeholders like Biochemists, Toxicologists, Sensory Scientists, Microbiologists, and Food scientists contribute in the backend for improving endurance in athletes indirectly. Moving forward, deep into the niche of Healthcare, Diet, and Nutrition. Healthcare covered the aspects of what different kinds of healthcare specialists are involved in an athlete's life, for example, Chiropractors, orthopedics, neurologists, it also included Dieticians and Nutritionists, from where it all started to explore how proper nutrition fuels performance and recovery.

Whether one tends to train for a marathon or is jogging to stay fit, diet plays a crucial role in fueling performance, aiding recovery, and maintaining overall health. This analysis approach involved five steps, as it started with identifying key nutrients for runners. Next, we listed down food sources for each nutrient determining the purpose and benefits. After that, recommend ideal intake levels and timings. Finally, potential challenges have been, such as travel or allergies related to Diet and Nutrition.

Further detailing the key nutrients for runners with their sources, which included carbohydrates to provide energy, proteins to support muscle repair, fats to regulate hormones and store energy, vitamins to ensure overall health, minerals to maintain bone and the functioning of nerves, and water, which is crucial for hydration.

Further detailing the key nutrients for runners with their sources, which included carbohydrates to provide energy, proteins to support muscle repair, fats to regulate hormones and store energy, vitamins to ensure overall health, minerals to maintain bone and the functioning of nerves, and water, which is crucial for hydration. All the whole grains such as brown rice, white rice, millets (bajra, jowar, ragi), wheat, oats, also Legumes like Lentils (toor dal, moong dal, masoor dal), chickpeas (chole), kidney beans (rajma), also fruits such as bananas, mangoes, apples, guava, papaya, oranges, pomegranate and vegetables like sweet potatoes, potatoes, carrots, peas, pumpkin has rich sources of carbohydrates which can serve as energy providers for prolonged activity as the recommended dosage is about 6–10 g/kg body weight per day. It serves as the primary fuel especially during prolonged exercise (90 + mins), as they are the primary sources of energy. Even required during the pre-event for about 1-4 g/kg body weight for about 3–4 h before exercise. Similar to carbohydrates, proteins, healthy fats, omega 3 fatty acids, calcium, fluids, vitamins and electrolytes were also mapped in higher detail.

Next, narrowed down to nutraceuticals and hydration, both of which are widely essential for the body but often misunderstood. Nutraceuticals enhance energy production, improve muscle recovery, and support cardiovascular health, while hydration prevents dehydration-related fatigue, regulates body temperature, and aids nutrient transport during intense physical activity [5]. By intersecting healthcare, diet and nutrition, with training to identify gaps in supplementation strategies. Then segregated nutraceuticals into traditional Indian remedies, such as Ashwagandha, Chawanprash, versus chemical-based ones, such as protein bars, caffeine, to understand cultural relevance and for the evaluation of their effects and side effects on athletes, and finally, analyzing different forms such as powder, capsules, and bars to assess their pros and cons.

Similarly, dehydration is a major issue not only due to the climate but also in the context of working out, training, and running, yet hydration strategies are rarely tailored to individual needs, where studying products like electrolyte drinks and homemade solutions [6]. In India's heat, hydration is non-negotiable, and dehydration can lead to muscle cramps or even heatstroke. Neglecting proper supplement use can lead to kidney damage or hormonal issues. Similarly, poor hydration doesn't just affect performance, but it also threatens health. Imagine collapsing mid-run due to heat exhaustion! Both scenarios are preventable with well-informed choices. The need to understand mental health as it is one of the key factors that impacts the athlete's ability to complete the routine. So, shifted the focus to do a ZIP analysis on mental well-being, which has received increasing attention in both academic and public discourse, especially following high-profile cases of elite athletes prioritizing their mental health nowadays. Often athletes prioritize performance over health, leading to burnout, depression, and chronic stress.

Communities provide emotional support, accountability, and a sense of belonging, which are the key factors for mental resilience. It also helped in understanding their role in fostering inclusivity, promoting grassroots development, and building support systems for athletes. An approach that began with focusing on Indian sports communities is Fig. 7.2. It involved analyzing case studies from urban and rural areas to capture diverse perspectives, also examining government policies supporting community sports, and review challenges faced by individuals. As part of the research, reviewing literature such as Sports Studies in India, analyzing case studies such as Mumbai's football clubs or Sherpas in Himalayan mountaineering, and examining policies of *Khelo India* to understand how communities help in enhancing endurance among athletes was done to improve the overall understanding.

Looking into an overall view, since policies shape the invisible backbone of sports systems, influencing athlete welfare and societal outcomes, the need to study different

Fig. 7.2 Various communities involved and key events

Government policies in the Indian context for promoting athletes' growth, which were mainly responsible for:

1. Providing a framework for sports development
2. Guiding resource allocation and infrastructure development
3. Promoting inclusivity and equal opportunities in sports
4. Enhancing India's competitiveness in international sports

This allows to identify persistent gaps that remain despite the country's development and still require attention. Some of the key policies Table 7.1 [7–13].

After understanding various policies, the focus shifted to studying Dress and Attire. Attire isn't just about looks; it's about comfort and performance. It also directly impacts identity formation and social outcomes. Athletic uniforms enhances athlete identity, which mediates mental health positively, and also reduces bullying by fostering equality among peers. The process involved five steps to understand this. First, review of existing literature on sports attire to understand its evolution. Next, analyzing case studies of innovative attire, such as moisture-wicking fabrics.

Table 7.1 Various key policies and details [7–13]

Key policy	Description	Impact on endurance	Governing body	Implementation region/ Demographics
National Sports Policy 2001	A policy by the Ministry of Youth Affairs and Sports to promote sports culture and integrate physical education into the educational curriculum	Encourages early involvement in sports, fostering physical fitness and endurance from a young age	Ministry of Youth Affairs and Sports, Government of India	Nationwide, Targeting educational institutions and sports organizations across India
Target Olympic Podium Scheme (TOPS)	Launched in 2014 to identify and support potential Olympic and Paralympic medal prospects with financial assistance for training and international exposure	Provides personalized support, enabling access to advanced training programs that improve athletes' endurance	Ministry of Youth Affairs and Sports, with support from the Sports Authority of India (SAI)	National initiative focusing on elite athletes across various sports disciplines
Special Area Games (SAG) Scheme	Initiated in 1986 to identify and train young talent from tribal, rural, hilly, and coastal regions with natural athletic potential	Provides specialized coaching to enhance the endurance and performance of athletes from these regions	Sports Authority of India (SAI)	Implemented in various states, focusing on athletes from tribal, rural, hilly, and coastal areas
Fit India Movement	Launched in 2019 to encourage citizens to include physical activities and sports in their daily lives, promoting a culture of fitness	Fosters regular physical activity, contributing to improved cardiovascular health and endurance among the population	Spearheaded by the Government of India, involving various ministries and stakeholders	Targets all citizens across India, promoting fitness activities in schools, workplaces, and communities
National Anti-Doping Agency	Established to promote, coordinate, and monitor the doping control program in sports across India	Enforces anti-doping regulations, ensuring athletes enhance endurance through legitimate training methods	Operates under the Ministry of Youth Affairs and Sports, Government of India	Covers all athletes and sports organizations nationwide

(continued)

Table 7.1 (continued)

Key policy	Description	Impact on endurance	Governing body	Implementation region/ Demographics
Khelo India Programme	Launched in 2017–18 to revitalize the sports culture at the grassroots level by building a strong framework for all sports and establishing India as a great sporting nation	Identifies young talent and provides quality training and facilities, enhancing the endurance of emerging athletes	Implemented by the Ministry of Youth Affairs and Sports, Government of India	Focuses on young athletes across various regions and demographics nationwide
Bharatiya Khel Initiative	An initiative to promote indigenous games and sports, preserving and encouraging traditional Indian sports	Encourages participation in traditional sports, contributing to overall physical fitness and endurance	Various organizations and government bodies promoting indigenous sports	Focuses on regions where traditional sports are prevalent, targeting local communities and youth
Scheme of Pension	Provides lifelong monthly pension to athletes who have won medals in international competitions, recognizing their contributions	Offers financial security, allowing athletes to focus on training and performance, thereby enhancing endurance	Ministry of Youth Affairs and Sports, Government of India	Applicable to eligible athletes across India who have achieved specified international accolades

Then, examining the psychological impact of attire on athletes. After that, assessed comfort and safety features like breathability and protection. Finally, evaluated how attire reflects brand identity and sponsorship.

Next, from intersecting various aspects of Training, Healthcare, Diet and Nutrition, Mental well-being, Communities, Policies, Attire, going into the ZIP analysis of Technology, in order to look at technology not as a simple tool but a game-changer in sports. The approach combined theoretical research with practical case studies as studying wearable technology like smartwatches and fitness trackers to understand their impact on performance metrics. Also examined existing AI systems that predict injuries based on player data. Additionally, looking at how social media and VR technology enhance fan engagement. All together shows that Technology enhances performance by providing real-time feedback, ensures safety by predicting injuries, and boosts fan engagement through immersive experiences like Virtual Reality. This integration of tech not only transforms how athletes train but also how fans interact with sports [14].

Following the daily life of an athlete, Skincare also came into focus on what an athlete is supposed to do throughout their day, be it pre-training, during training, or post-training. Skincare prevents skin injuries and infections, enhances recovery by reducing irritation, supports mental well-being through confidence in appearance, and maintains skin health under extreme conditions like intense sun exposure or cold weather. A five-step approach to understand this further, beginning with reviewing of existing literature on athletes' skincare to understand common challenges. Next, case studies analysis of athletes experiencing skin issues like chafing or sunburn. Then, skincare routines tailored to different sports conditions, such as high-altitude or humid environments. After that, assessing the impact of skincare on mental health, recognizing how healthy skin boosts confidence. Finally, evaluated skincare products specifically designed for athletes, considering factors like sweat resistance and hydration as Table 7.2.

7.4 Envisioning the Interventions

The Gigamap showcases the full complexity of the system by connecting all the different factors and illustrates the problem from a holistic perspective, enabling the identification of key systemic influencers, feedback loops, and intervention points for characterizing the system under consideration. The gaps between training, recovery, nutrition, and technology were addressed, enhancing the effectiveness of interventions and creating a unified framework to address challenges comprehensively.

Coming up with insights like Revenge effects that highlight the risks of overtraining, as what seems to be beneficial today can backfire tomorrow if recovery is neglected. Next, overtraining, which leads to fatigue, injury, and burnout, underscores the need for a balanced approach between pushing limits and allowing adequate recovery. Endurance is an emergent phenomenon, as it arises from the dynamic interactions between training, nutrition, healthcare, environment, and lifestyle. Each component contributes to overall resilience and performance, but it's their interdependencies that shape an athlete's ability to endure. Lifestyle supports daily function, which in turn influences training. Their training is shaped by nutrition and healthcare, which are themselves influenced by environmental factors. Applying these concepts to the project involved balancing training and recovery to avoid overtraining, which is the reverse effect, integrating all these components of the emergent system, and optimizing each nested system to enhance overall endurance.

Then to identify potential intervention points where targeted actions can have the greatest impact, also analyzing feedback loops to understand where small changes can have significant effects. Identifying key intervention points helps improve system efficiency, create lasting change by tackling root causes, and make better use of resources to achieve goals. The five intervention points, which discovered, included:

1. Training and rest schedule

Table 7.2 Factors, causes, effects, impact and solutions of skincare

Factors	Causes	Effects	Impact on endurance	Solution
Protection against friction and chafing	Repetitive motion, Sweat, tight clothing, Improper gear	Skin irritation, Rashes, Blisters, Painful raw skin	Chafing and friction-related injuries cause discomfort and pain, leading to slower performance and premature fatigue. Severe cases can force runners to stop or alter their gait, increasing energy expenditure	Anti-chafing balms, Powders, Compression gear, Compression gear, Seamless clothing
Hydration and moisture balance	Long exposure to the sun and wind dehydrates the skin	Dry, cracked skin, which can lead to infections and discomfort	Dehydrated skin can result in irritation, cracking and even minor wounds, increasing discomfort and making movement painful. Proper skin hydration ensures smoother motion and prevents unnecessary distractions	Hydrating lotions, Barrier creams, Electrolyte-infused skincare products
UV Protection and Sun damage prevention	Long hours in outdoor training under direct sunlight	Sunburns, Premature aging, Skin cancer risks, Heat exhaustion	Sunburned skin reduces the thermoregulation, making it harder for athletes to cool down, leading to dehydration and overheating, which impact stamina	Sweat-resistant sunscreens, UV-protective clothing, After-sun recovery products

(continued)

Table 7.2 (continued)

Factors	Causes	Effects	Impact on endurance	Solution
Sweat and Bacteria management	Prolonged sweating, Clogged pores, Microbial growth	Acne, Fungal infections (Athlete's Foot), Body Odor	Excessive sweating can lead to bacterial and fungal infections, causing discomfort and potential skin conditions that may force breaks in training	Antibacterial wipes, Sweat-wicking fabrics, Specialized cleansers
Blister and Foot care	Constant rubbing, Wet conditions, Ill-fitting Shoes	Painful blisters, Calluses, cracked heels	Blisters and foot pain alter a runner's stride, making movement inefficient and increasing the risk of injuries or knee pain. Healthy feet contribute to pain-free runs	Moisturizing foot creams, Anti-blister socks, Specialized bandages
Recovery and Skin repair	Exposure to harsh elements, Muscle fatigue, Prolonged exertion	Inflammation, skin breakdown, Delayed healing	Slow recovery leads to prolonged discomfort, affecting future training sessions. Faster healing ensures athletes can train consistently, improving overall endurance over time	Cooling gels, Recovery balms, Massage oils with anti-inflammatory properties
Cold weather skin protection	Dry air, Wind chill, Freezing temperatures	Frostbite, Redness, Cracked lips	Cold-induced skin damage can cause stiffness, numbness and reduced blood circulation, making it harder to maintain long distance running performance	UV-protective clothing

2. Mental well-being targeting sleep
3. Prevention, specifically Recovery
4. Nutrition, specifically hydration and
5. Technology highlighting personalization

Since several critical factors influence endurance, understanding how intervention points emerged was essential for optimizing performance. Several points that got highlighted included inadequate nutrition and hydration leading to dehydration and energy loss, while performance pressures impact mental well-being, causing sleep issues and anxiety, and overtraining without rest results in muscle damage and burnout. The absence of personalized training data leads to inconsistent training methods. Injuries disrupt routines and mental health, emphasizing the need for proactive prevention.

Once the intervention points were identified, it was now time to identify the leverage point, the touchpoint that affects endurance the most, showcasing the maximum impact with the least effort. This approach involved five steps: First, identifying potential intervention points. Next, the impact of each point on performance was evaluated as seen below in the Table 7.3. Then, the resource efficiency and challenges associated with each point was assessed. Justifying the selection of leverage points based on insights gained. Finally, developing strategies for implementing and monitoring these points. From the analysis, recovery is a critical leverage point due to its high impact on performance and injury prevention, offering immediate benefits with minimal effort. By focusing on prevention, athletes can reduce downtime and enhance overall resilience.

7.5 Formulating the Interventions

Once the Leverage point has been finalized, brainstorming of solutions based on the insights gained through various steps and mapping processes was followed to bring an overall holistic angle to the concepts. Solutions were ideated to address gaps in athlete recovery systems, enhance performance through innovative tools, promote awareness of sustainable endurance practices, and support athletes with tailored recovery strategies. Thus, the focus was on awareness, technology, and preventive measures for ideating solutions. The aim of which is to tackle critical recovery challenges by awareness campaigns, advanced technologies like sleeping pods, and wearable suit concepts.

Conducted impact analysis to evaluate recovery solutions for their effectiveness, to understand their impact on athlete performance, identify challenges, and determine best use cases. This ensures that resources are allocated efficiently to achieve maximum benefits. Critical analysis was conducted for proposed concept solutions such as the Athlete Suit, Sleeping Pod, and Social Media Campaign and the analysis revealed that the Athlete Suit concept has the highest impact on recovery due to its real-time fatigue monitoring and injury prevention capabilities, while The Sleeping

Table 7.3 Analysis of intervention points to find out the leverage point

Intervention point	Impact on performance	Resource efficiency	Challenges	Justification for leverage point
Training and rest schedule	Critical for preventing overtraining and ensuring optimal performance	Requires constant monitoring and precise adjustments	Hard to find the perfect balance and constantly adjust for individual needs	While vital, training/rest schedules need ongoing adjustments, making them less efficient than recovery
Mental well-being	Essential for motivation, Focus, and Mental resilience	Requires time for therapy, Mindfulness, and Consistent mental effort	Hard to measure progress and immediate impact on physical performance	Mental well-being is crucial, but its direct impact on physical endurance is less significant than recovery strategies
Nutrition (Hydration)	Supports energy production, Muscle repair, and Endurance	Relatively easy to implement; Significant benefits with proper hydration and diet	Needs consistency and education, but less immediate impact than recovery	Nutrition is vital, but without recovery, the body cannot fully utilize the benefits of good nutrition
Technology (Personalization)	Can optimize training, Hydration, and Sleep for better performance	High initial cost but continuous data feedback is helpful	Requires correct data interpretation and may overwhelm athletes	While helpful, tech needs to be paired with proper recovery for optimal results, making it less impactful on its own
Recovery (Leverage Point)	Directly influences muscle repair, Energy restoration, and Injury prevention	Highly efficient; Offers high returns with minimal effort (e.g., sleep, Stretching)	Requires time, Consistency, and Is often overlooked by athletes	Recovery has the highest impact on performance and injury prevention, is efficient, and provides immediate benefits

Pod enhances deep sleep but faces accessibility challenges and the Social Media Campaign promotes awareness but requires sustained engagement for long-term adoption.

Out of the three design concepts explored, ultimately choosing the Athlete suit as it had the most direct and meaningful impact on the athlete's day-to-day life. It wasn't just about adding smart tech, but it was also about enhancing how athletes train, recover, and stay safe while doing what they love. What really stood out was how this suit concept could support athletes during their training sessions, which is when most injuries tend to occur. Many athletes push themselves hard in training to improve their performance, but that often comes at the risk of muscle fatigue, strain, or even serious injury. And unfortunately, those training injuries can seriously affect their ability to perform in the future aligned matches or tournaments.

The suit concept is designed to bridge that gap by helping athletes train harder and smarter, without compromising their health. It acts as a kind of comfort shield, allowing them to focus fully on their workouts while knowing that the suit concept is monitoring their muscle health, regulating temperature, and even providing adaptive support when their body shows signs of fatigue or overuse. By offering this kind of real-time feedback and responsive support, the suit concept empowers athletes to push their limits more safely, which can lead to better long-term performance and fewer setbacks from preventable injuries. This concept wasn't just the most functional, but also has a futuristic touch and vision added to it. It meets athletes where they are, during the most intense parts of their routine, and works with their body, not against it, which is a major reason for consideration. After finalizing the idea of designing a suit, brainstorming sessions were held for the features that it can have, which can aid the athlete. To begin with, identifying the athlete's pain points during training, such as muscle stress and fatigue, muscle tear, and breaking of joints in extreme cases, were some serious considerations.

The goal was to prevent injuries proactively rather than reactively, enhance recovery efficiency, and promote safe training practices, which would eventually improve their endurance, focusing on temperature effects on efficiency and muscle health. In the Athlete Suit idea, the first goal was to brainstorm and visualize how wearable tech could actually benefit athletes in a real, practical way. So, began by creating a series of concept sketches for exploring different types of sensors and strap placements as seen below in Fig. 7.3.

Making sure that its not designing something that looks good, but something that could actually work with the athlete's body, movements, and routines. Compared these concepts by mapping them onto different athlete profiles, such as runners,

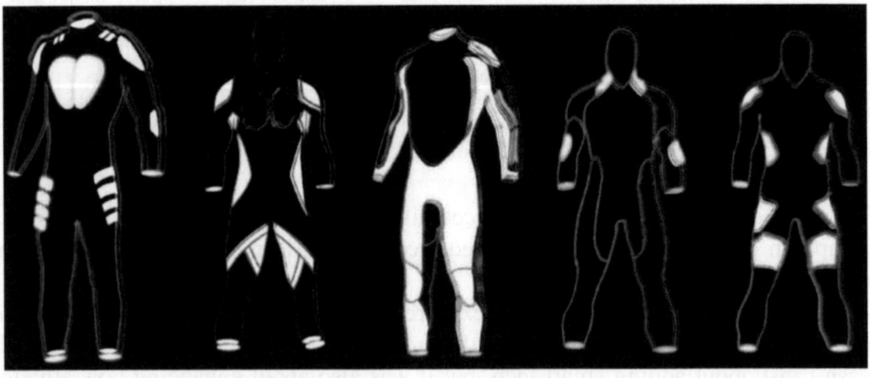

Fig. 7.3 Conceptual sketches, initial iterations of the concept

weightlifters, footballers, etc., to understand how the design would behave in actual use-case scenarios. After a series of iterations, finalizing on one concept that stood out as it had the best placement of straps and sensors that allowed for accurate tracking without sacrificing comfort, which was really important. If it's not comfortable, athletes won't wear it, no matter how advanced the tech is. But comfort and design were just one part of the equation. How will this actually help an athlete in their daily life? Do they consider this, especially in the Indian context? That's when the focus shifted on the core problem, which was injuries. As discussed, most injuries happen during training, not during performance. The idea was that "what if the suit could act as an early warning system?".

7.6 Enabling the Interventions

This final suit design concept (Fig. 7.4) tracks muscle activity in real-time. As it detects that a muscle is under strain and approaching a risky threshold, like the point where it might tear, then it sends an alert. This way, athletes get a heads-up before any serious damage happens. The idea is to help them push their limits, but safely, be it someone who is just starting out or training professionally. In short, we're not just building a wearable, but we're designing a training companion that can potentially extend an athlete's career by reducing preventable injuries.

This final concept was shaped by everything learnt through the extensive research and mapping analysis. Based on the data gathered and to identify the most effective placements for the sensors and straps, ensuring they aligned with the muscle groups most prone to strain or injury during training. The design concept wasn't just about where things could go, but where they would actually be useful without getting in the way of movement or comfort. The integrated sensors help detect any unusual

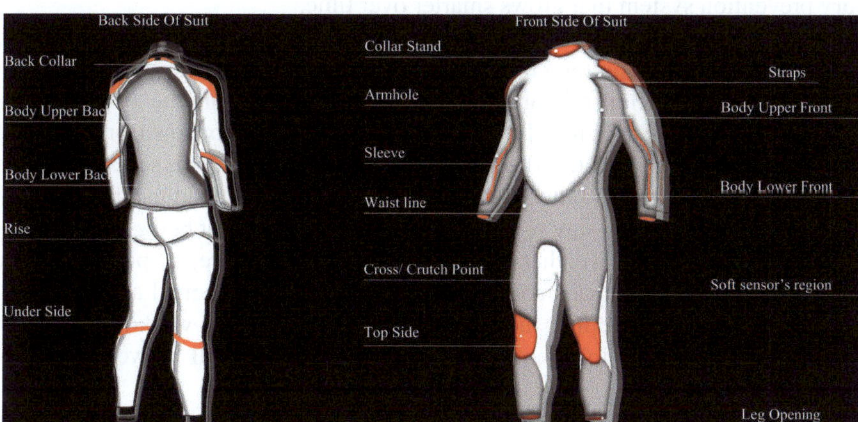

Fig. 7.4 Final conceptual design of the suit

strain patterns before they lead to injuries. Meanwhile, the straps are designed to hold everything securely in place, no matter how intense the workout is.

Finalization of this concept by balancing three key factors: functionality, sensor and fabric placement, and overall aesthetics making sure the suit design concept, apart from working well, needs to feel good to wear and look appealing enough that athletes would actually want to use it regularly. In the end, it's about creating a seamless experience where the athlete and the technology work together helping them train smarter, avoid injuries, and feel supported throughout their routine. Started by identifying key features that could make a real difference. One of the main focuses was on tracking muscle stress and offering adaptive support in real time through integrated fabric-embedded sensors that monitor muscle movement continuously. These sensors help detect patterns of strain, fatigue, or overuse. What's really smart about them is that they learn from the athlete's previous training data. That means, over time, the suit idea can identify when a muscle is being pushed beyond its usual limit and alert the athlete before it reaches a potential tear. Think of it like a warning system for the muscles, helping one train smarter and avoid injury, rather than going for rigid, robotic components. This ensures that the designed suit concept is much lighter and more flexible, which makes it feel like a second skin rather than a bulky wearable. The adaptive support in the suit functions almost like an artificial muscle, as it provides support where needed but still allows for natural movement.

One of the key considerations was the type of sensors used to track muscle health in real time. Simply gathering data wasn't enough, as it had to be accurate, responsive, and meaningful to the athlete. Soft sensors were proposed, specifically scar-based soft sensors. These sensors are incredibly sensitive to deformation, which means they can pick up subtle changes in the muscle as it contracts and stretches. What's even more powerful is that these sensors use historical feedback from previous training sessions to set a personalized threshold for each athlete. So instead of just measuring generic strain, the idea is to focus on what's "normal" for that individual and can issue alerts if the muscle is approaching a risky fatigue level. It's essentially a built-in injury prevention system that grows smarter over time.

Temperature control was another critical aspect. Training environments vary a lot, and overheating or feeling too cold can affect performance. Usage of temperature-regulating textile, which is a smart fabric that responds to the athlete's environment also supports the cause as in hot climates, they help cool the body down, and in colder conditions, they retain warmth to keep muscles flexible. The fabric is specially coated to enhance these properties, allowing the suit design concept to adjust dynamically. This means the athlete can stay focused on their training without being distracted by discomfort, no matter the weather. To manage body temperature, proposed adaptive yarn textiles, which are smart fabrics that respond to environmental and body heat changes. Whether it's keeping the athlete cool during intense workouts or warm during early morning training in cold climates, the fabric will actively adjust to support the body's needs.

Another critical component of the suit design concept is the conductive fabric used to design the support straps. These straps are conceptualized using a conductive textile embedded with shape memory alloy (SMA) springs. These alloys have a fascinating property: they "remember" their original shape and respond when heat is introduced. In this case, the heat doesn't need to come from an external source as it's generated naturally by the athlete's body, specifically during repetitive muscle movement. So the idea is, during training, the muscles are repeatedly activated, muscle temperature gradually increases. This localized rise in temperature activates the shape memory springs embedded within the strap's fabric. As the springs respond, the fabric contracts slightly, creating gentle tension that naturally restricts excessive muscle movement.

This contraction isn't meant to stop the athlete as it's designed to act like an extra layer of muscle support, almost like a smart brace that activates only when needed. It offers targeted resistance, helping the athlete train in challenging conditions while reducing the risk of overexertion or injury. Essentially, it serves as a safety net, especially in those moments where fatigue could lead to muscle tears or strain-related mishaps. And of course, comfort is just as important as function. That's the reason for conceptualizing the straps with sponge padding and breathable mesh fabric where they rest on the athlete's shoulders. This ensures that even with the smart support mechanisms in place, the suit design concept remains comfortable to wear over long sessions, preventing pressure points or discomfort. So, the straps aren't just holding the concept suit together, but they are working with the athlete, offering dynamic support when needed, and always keeping safety in mind.

The fabric material, which is one of the most innovative components of the suit concept. The idea was to cover the body, and to actively support the athlete by responding to their environment and physical state. The fabric is conceptualized using triacetate-cellulose bimorph fibers, with a thin layer of carbon nanotubes integrated into the weave. Now, what makes this fabric special is that it's built from metafiber yarns where each one is made by bundling textile fibers that have two key characteristics:

1. A conductive material embedded in just the right amount within the polymer base
2. An actuation mechanism that directly responds to changes in body temperature and skin humidity.

In simpler terms, this fabric will react to what's going on with the athlete's body. The envisioned functionality is at any normal resting state, let's say during warm-up or in cold weather, where the fabric remains less breathable. That's intentional. It helps retain body heat and keeps the athlete warm, which is important for maintaining muscle flexibility and preventing cold-related strain [15]. But as the athlete starts training, their body temperature rises, and they begin to sweat. That's when the real magic of the material kicks in, as the fibers are conceptualized as biomorphic, made of both hydrophilic cellulose (which absorbs water) and hydrophobic triacetate (which repels water), the sweat activates a coupling effect between the yarns. The fibers respond by subtly shifting, opening up the weave structure. As the yarns shift, the fabric's porosity increases, meaning it becomes more breathable right when the athlete needs it most. This increased breathability allows heat and moisture to escape more efficiently, helping to regulate body temperature and improve overall comfort.

What's especially exciting is that this all happens passively and in real-time. The suit concept doesn't require any switches, settings, or batteries to activate this function. The textile itself is conceptualized to adapt dynamically to the athlete's body, matching their physical output with thermal support as seen below in Fig. 7.5. So, in essence, the fabric isn't just clothing, it's a smart system that responds to the body's needs, helping the athlete perform at their best whether they're training in the cold or pushing through high-intensity workouts in warmer environments.

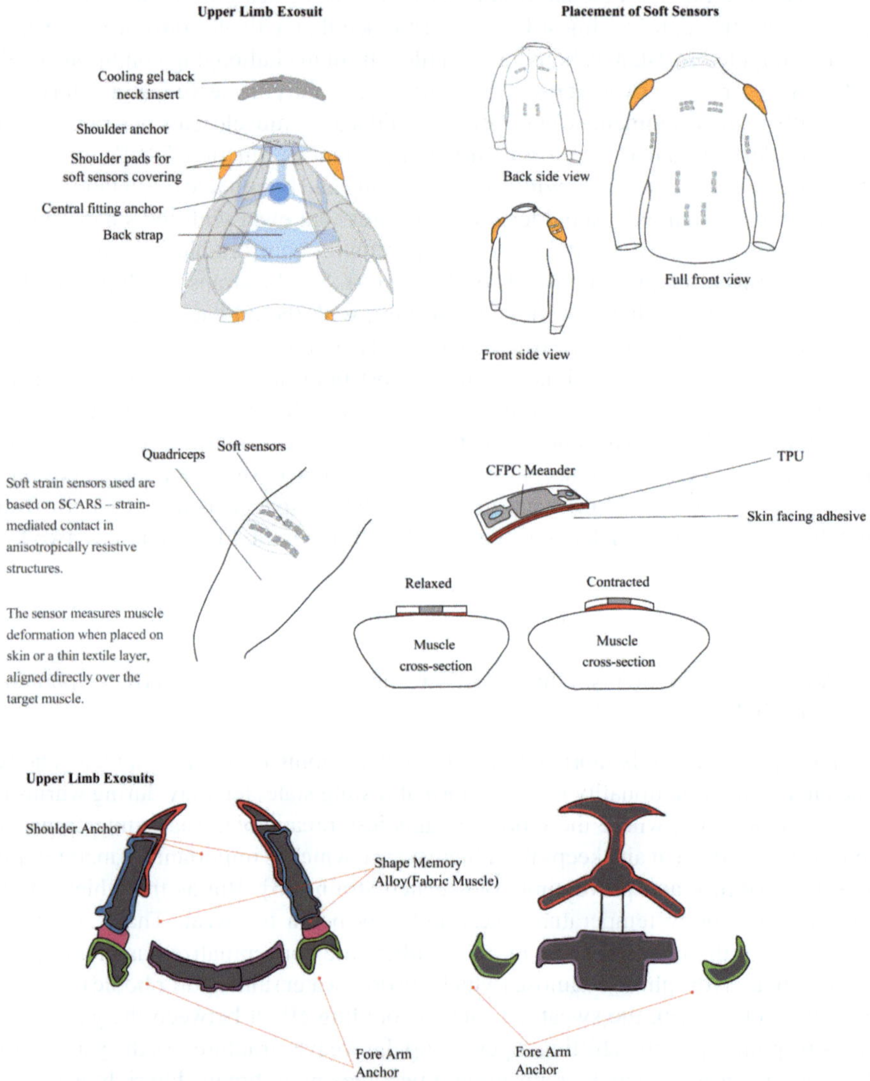

Fig. 7.5 Functional elements of the design with details

Finally, the entire design remains lightweight and flexible, so it doesn't restrict natural movement. Performance wear has to feel like part of the body and not a bulky add-on, which prioritized mobility and comfort without compromising on the designed suit's concept functionality. To sum it up, every element of the suit concept, whether it's the data-driven sensors, the adaptive support, or the intelligent fabric chosen and conceptualized, is to support athletes not just during performance, but during the crucial process of recovery and growth. It's like having a personal trainer, physical therapist, and smart assistant, all stitched into one unit. In the end, it's this combination of smart sensing, personalized feedback, and responsive textiles that makes the concept of a suit a real asset for athletes, helping them train better, recover faster, and avoid preventable injuries.

7.7 Conclusion

The systems design process explored key factors influencing endurance and proposed innovative solutions to optimize recovery and performance. The project began by understanding endurance, shaped by a multitude of factors including training, nutrition, healthcare, environment, and lifestyle. Through stakeholder mapping, the key influencers, such as coaches, institutions, and many others, were identified who directly or indirectly affect an Athlete's endurance.

The ideation focused on addressing athlete pain points, proposing solutions such as advanced sleeping pods, real-time monitoring suits, and awareness campaigns for sustainable recovery. The focus on functionality played a key role in formulating a design concept. This concept uses soft sensors to prevent injury, adaptive textiles to regulate body temperature, and user-friendly designs for flexibility. The main insight is that prevention in the sense of recovery is crucial for endurance. Proactive solutions and integrated tools like sensors, textiles, and campaigns can improve performance and long-term sustainability. By combining technology and awareness in the design concept, athletes can train safely, recover effectively, and perform at their best. This final solution in the form of a design concept was made possible by conceptualizing this issue as a systems design problem, with a focus on relationships between variables, emergent phenomena, and an adequate focus on obvious and non-obvious influences that finally shape the system!

References

1. Faga J, Bishop C, Maloney SJ (2023) Does size matter? Effects of small versus large pitch small-sided game training on speed and endurance in collegiate soccer players. Int J Sport Sci Coach 18(6):2033–2043
2. Scheer V, Tiller NB, Doutreleau S, Khodaee M, Knechtle B, Pasternak A, Rojas-Valverde D (2022) Potential long-term health problems associated with ultra-endurance running: a narrative review. Sport Med 52(4):725–740

3. Brach C (2017) The journey to become a health literate organization: a snapshot of health system improvement. Stud Health Technol Inform 240:203–237
4. Wang L, Kang Y, Wei L, Li M, Wang T (2024) Can the pitch dimension influence the physical fitness adaptations induced by small-sided training programs added to regular in-field training? A randomized controlled study in youth soccer players. J Sport Sci Med 23(2):487
5. Uthoff A, Oliver J, Cronin J, Harrison C, Winwood P (2018) A new direction to athletic performance: understanding the acute and longitudinal responses to backward running. Sport Med 48:1083–1096
6. Mantzios K, Ioannou LG, Panagiotaki Z, Ziaka S, Périard J D, Racinais S, Flouris AD (2021) Effects of weather parameters on endurance running performance: discipline-specific analysis of 1258 races. Med Sci Sport Exerc 54(1):153
7. Singh G, Bali A (2020) Play India youth games (Khelo India): scheme for promotion of sports in India. Int J Phys Educ, Sport Health 14(2):261
8. Pardeshi O (2025) India's future at the Olympics: a vision for athletic excellence. Research Nebula
9. Clarke J, Mondal S (2022) Sport policy in India. Int J Sport Policy Polit 14(4):729–741
10. Khatri K (2025) From policy to podium: evaluating the evolution and implementation of India's sports policies from 1984 to khelo India. Int J Phys Educ Sport Health 12:380–388
11. Star S (2023) The quest for harmonisation in anti-doping: an Indian perspective. Int Sport Law J 23(1):44–63
12. Singh A (2025) Sports law and policies: a study in Indian perspective. Indian J Leg Rev 5(7):166–171
13. Press Information Bureau (n.d.) Press Release. Government of India, New Delhi. Last accessed 16 Mar 2025
14. Burwitz L, Moore PM, Wilkinson DM (1994) Future directions for performance-related sports science research: An interdisciplinary approach. J Sport Sci 12(1):93–109
15. Esh CJ, Carter S, Galan-Lopez N, Garrandes F, Bermon S, Adami PE, Taylor L (2024) A review of elite athlete evidence-based knowledge and preparation for competing in the heat. J Sci Sport Exerc 6(3):218–237

Chapter 8
Improving Human-Canine Relationships through Empathy

Sayali Junagade, Sripriya Ganta, and Chawan Leela Krishna

Abstract The relationship between a dog and its owner is built on trust, communication and companionship, which is essential for their well-being. Strengthening this bond improves a dog's behaviour, emotional health and overall happiness while also enhancing the owner's companionship experience and responsibility. The primary aim of this project was to enhance dog-human relationships by considering the entire ecosystem in which they exist. We created a synthesis map from a detailed study of dog-human relationships from existing literature and presented an intervention in the form of a smart vest for health, behaviour, and emotional monitoring of the dog. We identified different stakeholders and focused on understanding dogs' behaviour, cognition, perceptions, life cycle and how they influence humans. We also examined how interactions between dogs and owners take place. Through Gigamapping, we focused on the key problems requiring intervention to create meaningful impact on the entire system. A leverage point with high potential opportunities and minimum intervention was identified by ideating solutions. After assessing costs and effects, we arrived at a less expensive solution that displayed a significant fit with the system. Through this process, we created a design concept of a smart vest for military and soldier dogs that improves communication, tracks performance and health monitoring, and helps owners better understand their dog's behaviour, strengthening the relationship between working dogs and their handlers. The insights from the gigamap indicate possibilities for future interventions in enhancing dog-human relationships.

Keywords Human canine relationship · Gigamapping · Human system relationship

S. Junagade (✉) · S. Ganta · C. L. Krishna
Indian Institute of Technology Kanpur, Kanpur, India
e-mail: sayalin24@iitk.ac.in

V. Kant (ed.), *Human Centred Design for Systems*, Springer Series in Design and Innovation 68, https://doi.org/10.1007/978-3-032-30376-9_8

8.1 Introduction

The bond between a dog and its owner isn't just about care; it's about connection, trust, and shared life experiences. But this relationship varies widely depending on who the owner is, whether a soldier, a farmer, an elderly person, or a child, and each bond has its own challenges. So, a systemic design approach to rethink the dog-owner relationship is needed in order to understand the interconnected and hidden connections. Using tools like Gigamaps and ZIP analysis, the ecosystem was mapped, key stakeholders were identified, and leverage points were uncovered, where small interventions could make a big difference. The ultimate goal was a concept design to be created for targeted, long-term solutions that support emotional, environmental, and social aspects of this relationship, not just as a quick fix, but a sustainable shift in the system. By understanding the unique needs of diverse owner groups, the focus is to build tools and systems that strengthen the human-canine bond in a meaningful, personalized way. It is crucial because when the relationship thrives, both dogs and humans thrive too in a mutually beneficial way.

8.2 Framing the System

The first significant step in framing the project was to recognize that dogs are companions, emotional support systems, and even working partners [1]. However, many dog owners, especially first-time adopters, face challenges in understanding and fulfilling their dog's physical, emotional, and behavioral needs. This gap in understanding often leads to stress, for both humans and the animal, strained relationships, and even abandonment in extreme cases. This topic aligns perfectly with following the human-centric systems design approach as it involves creating an ecosystem that includes awareness, communication, education, and service access, all of which require thoughtful design.

It can be considered as "a complex, emotional, and evolving journey; not a one-time design fix." Therefore, in approaching this issue, system design helps to take a holistic approach to emotional, environmental, and social factors. This helps to build long-term, integrated solutions rather than isolated interventions.

The goal was to improve the connection between a dog and its owner, a relationship that's built on trust, communication, and companionship. Strengthening this bond is essential because it directly impacts both the dog's well-being and the owner's experience. When this relationship is strong, it leads to better behavior, improved health, and a happier life for the dog. At the same time, it makes the owner's experience more fulfilling and meaningful, focusing on designing something that nurtures and supports the bond.

One of the main tools used was a Gigamap, which helped to visualize the entire system and understand how different elements are interconnected within the system. It reflects how everything in the system influences one another. A "ZIP analysis" was conducted, which helped to identify a leverage point within the system, a point where even a small intervention can lead to a significant change. These provided insights for creating the final design concept for improving human-canine relationships.

8.3 Comprehending the System

In order to understand the different stakeholders involved in the system and the kind of impact or influence they have in the system, stakeholder mapping was done categorising them based on the level of their impact into primary, secondary, and tertiary stakeholders. Primary stakeholders are people who are directly involved and have a strong influence on the system. This included different types of dog owners, such as soldiers, farmers, elderly people, children, etc. Understanding their needs, expectations, and feedback is crucial for the system to succeed. Secondary stakeholders were considered to be pet-product manufacturers, shelter providers, researchers, and breeders as these people may not directly interact with the dog-owner relationship every day, but could indirectly impact or influence the system. Lastly, in the tertiary category, groups like regulatory bodies and technology providers were identified, as they might not directly participate in the system but could be influenced by the outcomes of the project in the long run.

After completing the initial steps, we moved on to secondary research to build a deeper understanding of the subject. This phase focused on exploring several key areas critical to enhancing the human-dog relationship. We started by studying the different characteristics of dogs, including how breeds are categorized based on traits like size, temperament, energy levels, and functional roles (e.g, companion, working, guard dog, etc.). Simultaneously, the diverse profiles of dog owners, from first-time pet parents to experienced handlers, as well as people who adopt for emotional support, security, or companionship were also studied. In addition, we explored the life stages of dogs like puppyhood, adolescence, adulthood, and senior years to understand how needs and behaviors evolve.

In order to understand how dogs interpret the world, communicate with humans, and respond to their environments, studying their perception, cognition [2], and behavior [3] was essential. We recognized that "dog-owner relationship" isn't a monolithic concept as it varies significantly based on the owner's lifestyle and needs. Therefore, we captured this diversity, by categorizing owners into groups such as soldiers, farmers, herders, children, differently abled, elderly, and families, both rich and poor as each of these groups have unique interactions with their dogs. For example, a sheep herder relies on their dog for work, a child might see their dog as a playmate and companion, while a soldier's dog is a partner in high-stress situations. This ensured that the solutions would be relevant across these diverse scenarios.

Then, the key factors that influenced these relationships between people and their dogs were mapped out, which included:

- Breeds which give an understanding of suitability based on the owner's role and environment.
- Food and nutrition, where the diets are tailored to specific the dogs' lifestyles.
- Tech and equipment, where tools are used for training, assistance, or work efficiency.
- Emotions and behavior impacted by health, climate, and social interactions.
- Health, ensuring the dog's well-being.
- Training, essential for the dogs' behavior development.
- Shelter and living conditions, for providing appropriate environments.

A comprehensive matrix was developed to visualize how key factors such as training, nutrition, healthcare, and emotional support vary across different dog-owner groups. For instance, a soldier's dog demands rigorous discipline, specialized diets, and post-service emotional care, whereas a family pet might would require companionship, basic obedience, and play. This matrix helped to uncover the subtle yet significant differences in needs and expectations, enabling one to craft tailored interventions for each group.

In parallel, stakeholder mapping analysis was done to identify the roles of veterinarians, breeders, trainers, tech providers, animal welfare organizations, and government bodies. Their influence was examined and concomitant responsibilities, level of engagement, potential gaps, overlaps, and areas of synergy were identified. This dual-layered approach not only highlighted unmet needs but also revealed collaborative opportunities. It laid the groundwork for designing a more inclusive, sustainable, and responsive system that supports both dogs and their handlers throughout different life stages and circumstances. Understanding the dog's developmental phases is crucial for building a strong, healthy relationship. Each phase brings changes in skills, behavior, and needs. By studying these stages, training methods can be tailored, addressing behavioral issues, and ensuring overall well-being [4].

A few key areas were emphasized that shifts as the dogs grow. Firstly, food and nutrition, as a puppy, has very different dietary needs than an adult or senior dog. As they grow, their calorie intake, nutrient balance, and portion sizes all need to be adjusted to match their developmental stage. Next is training, as the techniques that are used for obedience and socialization need to evolve, too. Then there's physical growth, things like muscle development, body size, and strength. Looking at the mental and psychological development, as to how dogs perceive and react to the world around them. Their perception abilities were studied too on how the senses like vision, hearing, and smell, develop over time. Puppies are born with limited senses, and these senses develop quickly and become key tools for how they interact with their surroundings. Looking at behavioral changes in how social interactions, fears, and instincts evolve. For instance, a puppy may be playful and bold, while adolescent dogs often go through a rebellious or anxious phase, and older dogs may become more set in their ways or even show signs of separation anxiety. Finally, focusing on health needs, from timely vaccinations in early life to regular vet checkups, balanced exercise, and age-specific care as they grow older. These health considerations shift just as much as everything else does. So, overall understanding these developmental phases helps to create a more supportive environment for dogs, one that respects their needs and adapts as they grow.

This understanding of developmental phases might impact a specific owner category (Fig. 8.1). For example, a family with young children would require an understanding of the puppy's socialization phase. They would need to teach the puppy how to interact safely with children, while also ensuring that the children understand how to handle the puppy gently. Likewise, a herder would need to know when to start specific training for working dogs, which might differ from the training of a family pet. This detailed analysis of owner categories, influencing factors, and developmental phases would serve as a foundation for the solutions. Understanding the complexities of the dog-owner relationship helps to develop more effective and tailored interventions.

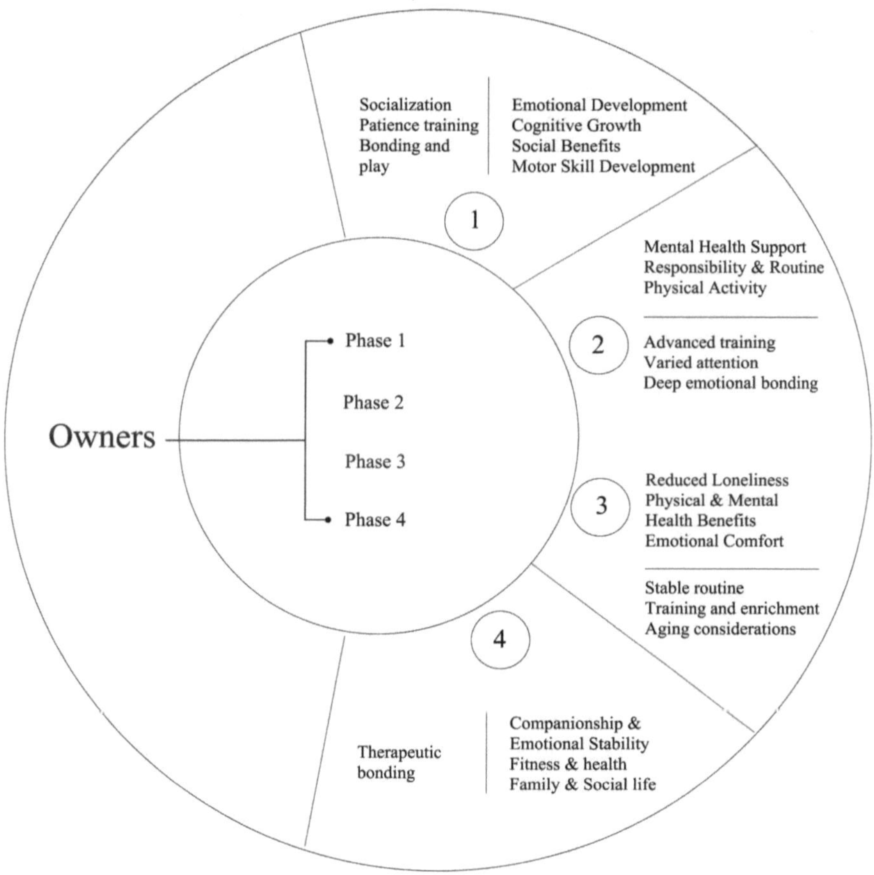

Fig. 8.1 Phases of owner dog bonding

The four stages of human life: childhood, teenage, adulthood, and old age, were mapped out and connected with pet dogs and the development of human-dog relationships. In childhood, dogs contribute to the child's emotional development, cognitive growth, motor skills, and social interaction. Kids not only learn about empathy but also develop strong interpersonal skills through that bond. Teenagers experience emotional shifts, and a dog can provide mental health support, build a sense of responsibility, encourage routine and physical activity, which often gets overlooked in this age group. In adulthood, dogs become a source of emotional comfort, help tackle loneliness, and offer significant physical and mental health benefits, especially considering the stress and fast pace of adult life. For older adults, dogs offer companionship, emotional stability, and help them stay physically active. They also contribute to maintaining a healthy social and family life, which is important in avoiding isolation. It's a dynamic relationship that adapts to the individual's needs. This approach shows how the life course of an individual shares generic dimensions with the human species and how these dimensions can be tapped into for revealing the influences that affect the dog-human relationship. It is interesting to understand human stages and how they influence each other.

Dating back to history [5], starting from the Prehistoric Era, wild wolves started scavenging near early human settlements, and over time, the less aggressive wolves began sticking around, forming a symbiotic relationship with humans. These wolves acted as early warning systems, alerting humans to dangers like predators or intruders. In the Domestication Phase, humans began feeding and sheltering more social wolves, where this mutual bond grew stronger as these early dogs played a vital role in hunting, tracking, and protection. Their loyalty and skills made them indispensable companions to early humans. By the time of the Agricultural Revolution around 12,000 BCE, dogs took on even more functional roles as humans transitioned from hunter-gatherers to settled agricultural societies. Dogs became essential for guarding livestock, herding animals, and controlling pests. The growing diversity of tasks led to the development of different breeds, each specialized for a particular function or terrain. This marked the beginning of a more structured relationship between humans and dogs, which continued to evolve as both species adapted to each other's needs.

In Ancient Civilizations spanning around 5000 BCE–500 CE in places like Egypt, Greece, Rome, and China, dogs were used for companionship, guarding, hunting, and even warfare. In the Medieval Period, dogs were seen in castles, farms, monasteries, and some specific breeds, like St. Bernards, were even trained for rescue missions. In the Industrial Period, Kennel Clubs formed an organization that concerned itself with the breeding, showing, and promotion of more than one breed of dog. Breeds were standardized, and dogs started entering middle-class homes as companions. And today, in the Modern Era, dogs support therapy, rescue work, emotional care, and even participate in scientific studies. Other people recognize that dogs process human emotions in similar ways, showing how deep and evolved this bond has become.

So far, dog development, the evolution of their relationships, and their impacts on owners were all discussed. Now, let's focus on how dogs perceive information or any given stimuli and how it enhances communication between dogs and humans. Dogs perceive the world through a rich and unique sensory lens that differs significantly from humans. Their olfactory system is incredibly advanced, enabling them to detect and interpret scent-based information far beyond human capabilities. This acute sense of smell allows dogs to perceive emotional changes in humans, such as stress or fear, through chemical cues in sweat, even before one become consciously aware of them. This deepens their empathetic connection and enhances their role as emotional companions.

In addition to scent, dogs have highly sensitive hearing. They can detect ultrasonic sounds and hear frequencies beyond the human range, which makes them adept at sensing potential danger and responding quickly to vocal cues or commands.

Their vision, while not as colour-rich as ours, excels in detecting motion and functioning in low-light conditions, giving them strong night vision. All of these combined sensory abilities offer dogs a multi-dimensional understanding of their environment, shaping how they behave, respond, and bond with humans.

By understanding a dog's superior senses (Fig. 8.2), communication in training and daily interactions can further be improved. For example, dogs have an exceptional sense of smell, which can be utilized in scent-based commands or scent-tracking activities. Their keen hearing allows them to respond to specific vocal tones, so using clear, consistent commands in a calm, assertive tone can enhance their understanding. Additionally, dogs are highly responsive to visual cues, such as hand signals or body language. By tailoring the communication to their sensory strengths, fostering better interactions, reinforce training, and strengthen the bond between humans and dogs. In therapy, for example, a dog's ability to sense anxiety through smell and respond with calming behavior is invaluable. Understanding this can allow therapists to maximize the dog's impact. Similarly, for a search and rescue dog, their olfactory senses are the key to tracking and finding people. It's about leveraging their strengths to bridge the communication gap. Dogs' unique perception abilities make them exceptional communicators in their own way. By recognizing how they sense and express emotions which can be used to deepen human connection with them. This understanding fosters trust, empathy, and a more meaningful relationship between humans and dogs, enhancing both companionship and overall well-being.

In order to understand how dogs interpret their world, dog cognition [2] has been explored, focusing on their "Theory of Mind", the ability to perceive human intentions and emotions. This involved studying how dogs learn through association, observation, and reinforcement, and how their memory works, especially in recalling routines and emotional experiences. This helped understand their behavior and emotional responses more deeply. Through Dog's Theory of Mind, understanding how dogs interpret emotions, follow human gestures and gaze, and understand intentions, though they still lack full self-awareness can be seen.

Their memory and learning abilities are also quite interesting as they show signs of:

- Episodic-like memory, remembering specific events
- Associative learning, connecting actions with rewards or punishments
- Observational learning and even

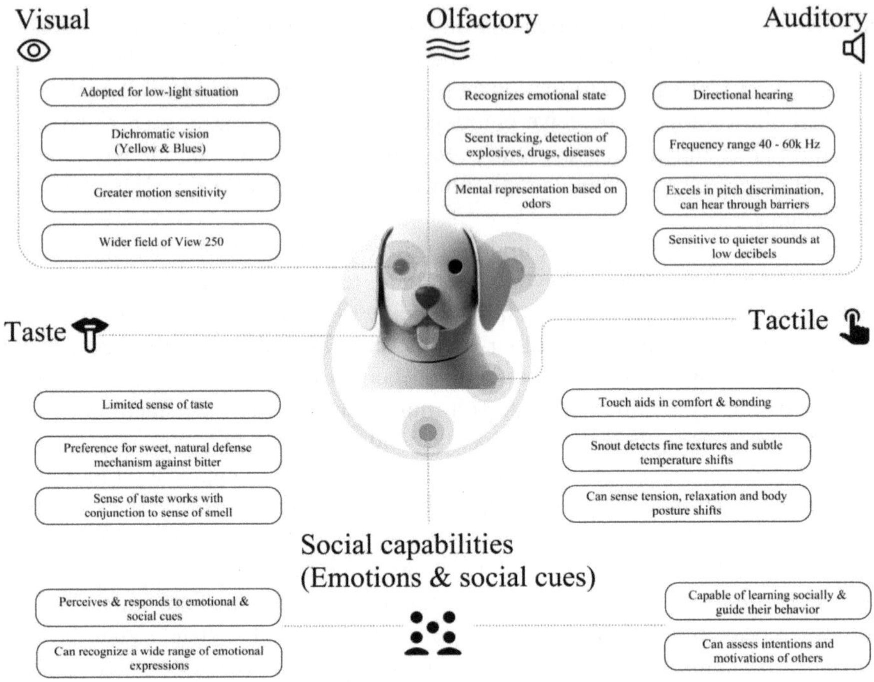

Fig. 8.2 Dog's perception

- Deferred imitation, where they copy actions they've seen earlier

Analyzing this from three perspectives, evolutionary, ecological, and anthropogenic, showing how dogs evolved from wolves, adapted to human environments, and developed social intelligence through domestication. They're strong in teamwork, social bonding, and reading human cues, but not as great at independent problem-solving.

Overall, dogs are emotionally intelligent and socially aware, with their behavior deeply shaped by human interaction. To understand them better, different types of behaviors they exhibit were studied, looking at playfulness, aggression, anxiety, or submission, and how they communicate emotions through body language, vocalizations, tail movement, and facial expressions, allowing them to express needs, stress, and affection.

Many owners don't realize how much a dog's posture and body language reveal their emotional state. Dogs communicate a wide range of feelings, like fear, anxiety, love, excitement, or submission, through subtle cues such as facial expressions, tail position, ear movement, and overall body posture (Tables 8.1, 8.2, 8.3 and 8.4).

For example, a crouched body with a tucked tail may indicate fear, while a relaxed body and soft eyes show comfort and affection. Even behaviors like yawning or lip licking can signal stress. By learning to read these signs, owners can better understand and respond to their dogs' emotions, strengthening the human-animal bond. Dogs express their emotions through various body movements, making it essential

Table 8.1 Dog's expressions and emotions–eyes

	Expression	Emotion
Eyes	Slow blinking	Affection and relaxation
	Darting eyes	Nervousness, fear, or uncertainty
	Whale eye	Anxiety, fear, or discomfort
	Dilated pupils	Excitement, fear, or arousal
	Hard stare	Challenge, dominance, or warning
	Narrowed eyes	Pain, discomfort, or submission
	Avoiding eye contact	Submission, fear, or respect

Table 8.2 Dog's expressions and emotions–ears

	Expression	Emotion
Ears	Relaxed	Calmness, contentment
	Rapid movement	Uncertainty, assessing the situation
	1 up,1 down	Confusion, curiosity, playfulness
	Flicking	Listening to multiple sounds, nervousness
	Pinned Back	Submission, friendliness, uncertainty
	Forward	Alertness, curiosity, excitement
	Relaxed	Calmness, contentment

Table 8.3 Dog's expressions and emotions–head

	Expression	Emotion
Head	Head jerk backwards	Surprise, dislike, or rejection
	Head tilt	Curiosity, attention, confusion, or trying understand something
	Head lowered	Submission, guilt, sadness, or appeasement
	Head held high and stiff	Confidence, dominance, alertness, or challenge
	Quick head turn away	Avoidance, discomfort, or submission
	Slow head turn	Interest, consideration, or mild suspicion
	Head shake	Stress release, discomfort, or ear irritation

Table 8.4 Dog's expressions and emotions–tail

	Expression	Emotion
Tail	Fast wagging	Excitement, happiness, or high energy
	Slow wagging	Caution, insecurity, or a calm, friendly greeting
	Wagging more to the right	Positive feelings, comfort, or recognition of a familiar person
	Wagging more to the left	Anxiety, uncertainty, or a potential threat
	Low but wagging	Submission or nervous friendliness
	Tucked between legs	Fear, submission, or anxiety
	Twitching or flicking tail	Irritation, minor annoyance, or slight aggression

to observe their tail, ears, heads, and eyes to understand their mood. These subtle cues offer valuable insights into their emotional state.

For instance, the tail is a strong indicator. A high, stiff tail often signals alertness, excitement, or dominance, whereas a low, tucked tail typically reflects fear, submission, or anxiety. Similarly, eye contact, ear positioning, and head tilt convey curiosity, discomfort, or attentiveness. By paying close attention to these non-verbal signals, humans can interpret a dog's needs more accurately, enabling more compassionate care and effective communication.

Dogs can also be categorized based on their roles and functions: companion dogs, service dogs, guard dogs, and therapy dogs, amongst others. When owners with such different requirements interact with various categories of dogs, unique patterns of interaction emerge. These interactions give insights into what are the key areas in human-dog relationships growing around.

Service Dogs are the dogs that are highly trained to assist individuals who are differently abled. They exhibit strong discipline, intelligence, and calmness. The owner relies on the dog for essential tasks, and the dog, in turn, is highly responsive. These owners are typically structured, disciplined, and deeply trust their dogs. These dogs are trained, and there are different techniques of training that are required. The most common technique used is through verbal commands and voice cues, using short words and tone to guide their actions. Along with that, hand signals and body language help dogs understand better, especially when paired with verbal cues. Clicker training is another technique that uses a clicking sound to mark the right behavior, followed by a reward. Positive reinforcement training, Luring, and shaping, socialization training, leash, and recall training were all available. Today, modern technology-based training tools, are also available:

- Smart collars
- AI bark translators
- Clicker training apps
- Communication buttons, where dogs are trained to press buttons to talk.

These training and techniques, in some or other way, help in building the bond. But some important components are directly involved in this relationship. Analysing key components that define their bond, such as retention rate, time spent together, emotional intensity, and dependency. Identifying extremities of components involved helps to analyse how bond building occurs and what the nature of that bond is. Focusing on measurable components in the dog owner relationship. For example, for Military Personnel, the components like strong bond through eye contact frequency, command retention, and high-stakes situations are some of the key components to be listed. For children, key components might change, such as bonding through play, eye contact, emotional response, and proximity.

Through all the analysis and research, the actual relationships and interconnections that existed within the system are well understood, which brought the system Gigamp together as one complete whole (Fig. 8.3). As the gigamap served as a visual representation of the research done on the human-dog relationship, it further helped to understand multiple layers of data and their connections. The central portion of

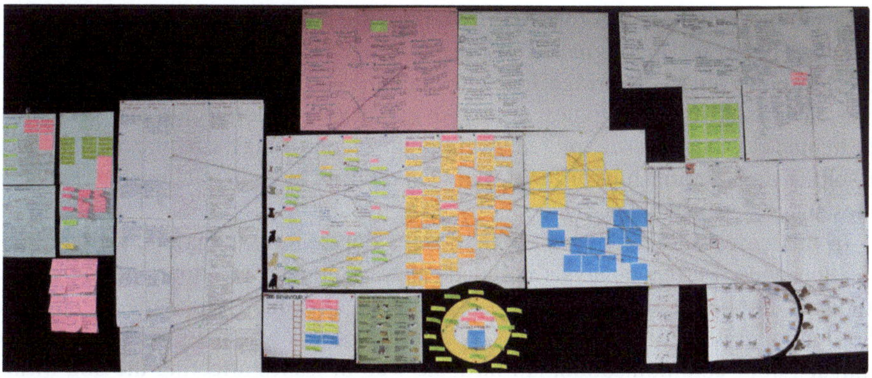

Fig. 8.3 Gigamap with interconnections

the gigamap highlights core themes such as dog development phases, its cognition, behavior, and perception. The connections made between the various components of the gigamap acted as a tool highlighting the interdependencies, identifying the gaps and opportunities, improving human-dog interactions.

8.4 Envisioning Intervention

The intervention points were then identified, as these points acted as those areas that had the power to alter the functioning of the system, either by opening up opportunities or addressing a problem. In order to understand the scope of change it has on the system, there is a need to detail the intervention point. ZIP analysis allowed for the systematic breakdown of problems and potentials in different dog-owner relationships. It helped to pinpoint key areas for "improvement" that are the ZIP points.

Soldiers and their dogs experiencing emotional distress due to long deployments were identified. For differently abled people, training gaps and accessibility issues hinder service dog effectiveness. Isolation and separation anxiety lead to destructive behavior. Abusive owners cause trauma, impacting socialization. Rapid adolescent changes challenge training and bonding too [6]. The key idea is that these ZIP points highlight specific challenges and potential interventions. By zooming in on these areas, targeted interventions that enhance companionship, security, and overall well-being can be created for. This leads to healthier, more balanced relationships. In essence, we're focusing on specific, actionable areas to improve the dog-owner bond. By addressing these challenges, impactful solutions can be created that benefit both dogs and owners.

Differently abled owners were categorized into mentally and physically disabled groups, including blind, deaf, mute, and those with locomotor impairments. Each group faces unique communication challenges as perception gaps play a critical role.

For example, blind owners rely heavily on auditory and tactile cues, whereas deaf owners depend on visual signals. These differences can lead to misunderstandings and affect responsiveness. To bridge these gaps, adaptive training methods, sensory-based cues, and assistive technologies, aiming to enhance communication and strengthen the human-dog bond were explored. Additionally, identifying the nature of challenges in soldier-dog relationships differed significantly from those in relationships between dogs and differently abled owners.

In this case, soldiers dogs face challenges due to stress, anxiety, and communication gaps. Further analysis based on these contexts was done, noting how a soldier's emotional state could impact the dog. The main problems found were the dogs inability to control their stress and anxiety, which is worsened by the human's poor communication. An intricate web of issues affecting the dog-owner relationship was created by the combination of soldier's diverse emotional states and severe operational demands. To mitigate these issues and foster a more effective partnership, structured training protocols along with the implementation of emotional support techniques for both soldiers and dog, and a heightened awareness among handlers regarding their emotional impact on their animals was focused on. Ultimately, the aim was to address the unique needs of these dedicated working dogs and their owners, thereby strengthening their relationship and enhancing their overall operational efficiency.

The effects of abuse on a dog's physical and mental health, including fear, anxiety, aggression, and trust issues were also studied. Abusive owners often appear as two types: Unintentional abusers and Intentional abusers. Unintentional abusers are the ones who may become abusive due to stress, financial issues, or lack of awareness. Education and awareness can aid in preventing this. Intentional abusers are the ones who abuse but do not know the deeper effects of it.

Also, a dog's adolescence phase, around 9 to 12 months, is critical just like the teenage years in humans. They go through hormonal, behavioral, and emotional changes. Male dogs may show humping or aggression while females begin their heat cycle, which brings physical and emotional shifts. Without proper care at this stage, they may become disobedient, anxious, or insecure. Hence, consistent training, socialization, and a safe environment are important considerations. Another issue is separation anxiety in dogs, which is something that affects around 20–40% of dogs globally, especially certain breeds like Beagles, Dachshunds, and Labrador Retrievers. It typically starts showing up when dogs are around 2 to 4 years old. There are many possible causes for it, as it could be due to the loss of a family member, moving to a new place, or a sudden change in ownership. All of this can trigger fear and anxiety in dogs when they're left alone. In terms of behavior, dogs may start showing signs such as excessive barking, whining, destructive chewing, urinating indoors, pacing, escape attempts, or even refusing food. Now, depending on how severe anxiety is, the treatment varies. For mild cases, even playing soft music or recordings of familiar voices can help the dog feel less alone. But in severe cases, teaching the dog to associate being alone with positive experiences is beneficial. This must be done gradually and carefully, step by step. So, the idea here is to create intervention methods such as interactive toys for dogs, along with, awareness and tech-based solutions or training programs.

8.5 Formulating the Interventions

A leverage point is an area where targeted interventions yield high impact for under-addressed issues with significant potential. With the goal of addressing current concerns, which when fixed, would result in significant improvements in behavior and relationships. Four main points were identified: Behavioral issues in the adolescent phase, dogs owned by soldiers and differently abled owners, dogs in healthcare, and dogs owned by abusive owners.

Moving onto ideating solutions for each, addressing their problems, and analyzing the solutions. After careful analysis, two strategic solutions were selected moving forward, an augmented reality app concept for behavior detection to suggestions and a smart vest concept for real-time emotion, stress, and health monitoring for a dog's emotional and physical state.

The augmented reality app is designed especially for new dog owners or individuals seeking to better understand their dog's behavior. It scans and detects current actions, analyzes them, and offers immediate suggestions. With built-in health monitoring and multiple functional modes, the vest ensures constant awareness of the dog's well-being. Further a cost-benefit analysis and impact assessment was conducted that evaluated the features, feasibility, and influence on communication.

The two key stakeholder groups were analyzed which are as follows:

- Differently-abled Owners
- Dogs Owned by Soldiers

For identifying key stakeholders, the different needs of each stakeholder were analyzed, and the areas were identified such that the needs can be holistically and effectively addressed with the product idea conceptualized. Then, the dogs owned by soldiers were chosen as the major emphasis point. The key idea was to systematically rate each idea based on effectiveness, affordability, and potential reach, to find the highest positive impact. Here, both the solutions arrived at were analyzed based on a few aspects to measure the effectiveness of the solution. Through this analysis, the most effective intervention was finalized, ensuring it would deliver meaningful improvements to the dog-owner relationship. Thus, the goals were to improve the efficiency and success rate of dog training and introduce better methods for dog-human interactions. While these relationships are being addressed, the key idea is to systematically focus on those aspects where interventions could be done and then propose solutions that are feasible and sustainable in the longer run.

The problem formulation and the solution go hand in hand. The effectiveness parameters used involved gesture and movement interpretation, Audio data, health monitoring, analysis of health data, and analysis of behavioral data to understand the feasibility of the product. The goal was to evaluate their feasibility based on cost

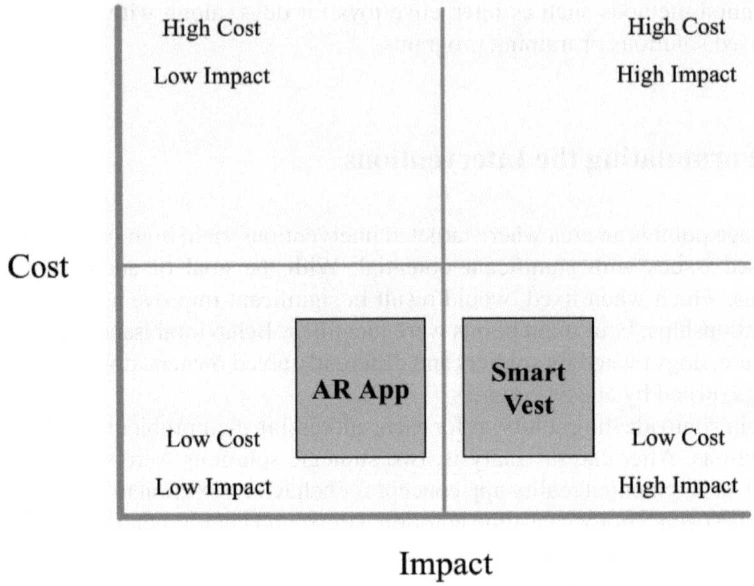

Fig. 8.4 Cost impact analysis

and impact. In Impact vs Cost Analysis (Fig. 8.4), for impact, the potential user base was considered, as to how far the product has the potential to reach, and what certain things the product can enhance through rating also considering the cost-benefit ratio. After analyzing the above for both solutions, "Smart Vests" was identified as the solution that can create a greater impact on the whole system with less cost and less effort.

These insights led to the development of the final design solution–"Chitta," a smart vest, which is an advanced wearable system that is designed to enhance communication, safety, and overall well-being of service dogs that basically helps both for assisting and even for training purposes. It enables the owners to understand and respond to their dog's needs effectively. Chitta means "emotional or feeling component of the mind", which ties perfectly to the goal that was envisioned, which was to improve the dog's behavior, health, and overall happiness while also enhancing the owner's experience.

In order to tailor the design solution as per user needs involved, the targeted user type was categorized into three key roles (Table 8.5):

- Handlers
- Veterinary Officers
- The Chief Training Officer

Handlers are the people who work directly with military dogs on missions like border security, explosives detection, or search and rescue. Their goal is to build a strong bond with their dogs, ensure they perform well during high-risk operations, and maintain the dogs health and discipline. What they really need is real-time health and stress tracking tools, and better communication systems during missions. Their biggest pain point is that they often have limited visibility into what their dog is going through, like stress or health issues. In addition, handling dogs in tough weather or terrain can be a real challenge, and emotional attachment makes it even harder.

Table 8.5 Target users

Types	Handlers	Veterinary officers	Chief training officer
Bio	Works directly with military dog he has been assigned Trained at the RVC Centre Usually assigned to border security, explosives detection or search & rescue	"Specializes in military canine health, surgery and rehabilitation Stationed at Army veterinary hospitals & training schools"	Senior commanding officer in charge of all dog training programs. Oversees RVC's strategy for breeding, training and mission deployment Directs innovation and policy improvements
Goals	Build a strong bond with their assigned dog Ensure the dog performs tasks effectively in high-risk missions Maintain the dog's health, fitness and discipline	Prevent injuries & long-term health issues in military dogs Improve nutrition and fitness plans for dogs Ensure medical readiness for deployment	Improve the efficiency and success rate of military dog training Introduce new ways for better canine management Ensure handlers and trainers receive the best methodologies
Needs	Lightweight, durable gear that allows free movement Real-time health & stress tracking of dogs Communication tools for mission updates	Wearable health-monitoring tech for early risk detection Detailed behavioral logs Real-time updates on dogs' vital signs	Research-based improvements to training techniques Insights into dog behavior
Pain points	Limited visibility into the dog's stress or health conditions Diagnosing stress, fatigue or dehydration Difficulty in handling dogs during extreme weather or terrain conditions High emotional attachment	Diagnosing stress, fatigue or dehydration in high-risk zones Delayed reporting of injuries during missions Limited data on long-term impacts of military service on dogs	Need for scalable solutions to train more dogs effectively Balancing operational effectiveness with the Well-being of dogs

The proposed solution can also help Veterinary Officers who specialize in the health, surgery, and rehabilitation of military dogs that are based at Army Veterinary Hospitals and Training Schools. Their focus is on preventing injuries and long-term health issues, improving nutrition and fitness, and making sure dogs are medically ready for deployment. They need detailed behavioral logs and real-time updates on a dog's vitals. But they struggle with diagnosing stress or fatigue in high-risk zones, delayed reporting of injuries, and a lack of long-term data on how military service impacts dogs over time.

And finally, the Chief Training Officer, who is the senior commanding officer who oversees all dog training programs. They're in charge of setting strategies for training, breeding, and mission deployment, and they drive innovation and policy updates. Their goals are to improve the efficiency and success rate of dog training and introduce better management methods.

The smart vest is designed to strengthen the bond between soldiers and their service dogs. It provides real-time monitoring of the dog's behavior, emotions, and health. It uses advanced sensors to track stress, body language, and vital health indicators.

The core features includes:

- Behavior & Emotion Tracking: Detects mood shifts, stress levels, and body language, allowing handlers to recognize anxiety or discomfort.
- Health Monitoring System: Measures heart rate, respiration, and stress indicators, providing long-term health trends.
- Proximity & Activity Analysis: Alerts handlers to fatigue, unusual behavior, or breaches of designated zones.
- Smart Temperature Regulation: Adjusts the vest's temperature for comfort in extreme environments.
- Behavior Tracking & Insights: Compiles data to understand patterns in the dog's stress and well-being.
- Time Spent Together Tracking: Promotes stronger emotional connections through interaction data.

In the initial design phase, various functional elements such as adjustable straps for a better fit, pockets for storage, and air circulation vents to ensure comfort were all explored (Fig. 8.5). Features such as touch sensors, strings that can tighten were also considered while designing. But the final version eliminates unnecessary bulk while maintaining key features, as the final version is designed to enhance both comfort and functionality for the dog, it was conceptualised to be made from Nylon PU-coated fabric for the outer layer, making it durable and weather-resistant, while the inner layer will be made of soft cotton for comfort. The design has integrated adjustable straps to ensure a snug yet comfortable fit for different dog sizes. A leash hoop for easy attachment, reducing the need for an extra harness. The reflective strips improve night-time visibility, enhancing safety. The dedicated sensor placements to monitor heart rate, respiration, and temperature help in tracking the dog's well-being.

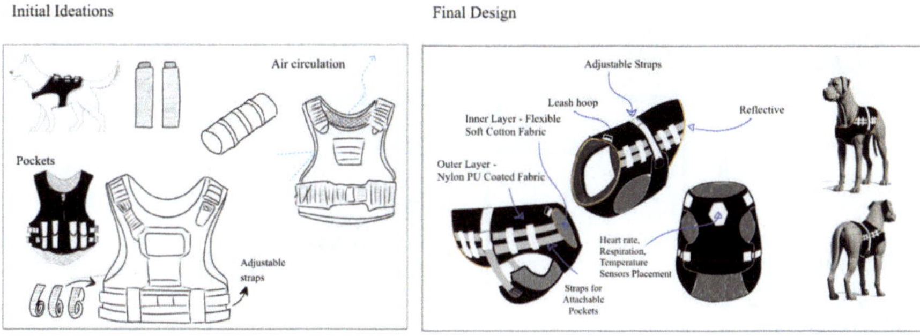

Fig. 8.5 Design concept

8.6 Enabling the Transition

The data collected by the vest is continuously synced with a mobile app, allowing owners to monitor their dog's condition in real time. This helps them stay informed about behavioral patterns, stress levels, and health indicators. In order to design the app interface, beginning with several iterations as the goal was to provide a high-level overview of the dog's well-being while also allowing users to explore more detailed data when needed. A home screen was envisioned which functions as a dashboard, guiding users to different components such as health, behavior, alerts, and progress tracking.

These ideas were translated into low-fidelity wireframes to map out the structure and navigation flow. These wireframes represent early-stage exploration of how the system could support intuitive, data-driven interaction between the user and their dog. Looking at the home screen, the topmost part provides data about dogs and adjustable modes. Next, there is an overview of the dog's health and behavioral data at the lower part. Users get immediate alerts, highlights, and recent recommendations. The navigation bar at the bottom helps to have a detailed view of each component's health, behavior, and vest status. The health monitoring screen displays key physical attributes like height, weight, age, and medical history of the dog (Fig. 8.6).

Let's consider some of the behavior monitoring and vest status tracking features of the app, as it has continuous tracking of current mood and stress based on behavioral signals, which are continuously monitored. The vest includes a built-in speaker to deliver audio commands. It also includes vibrational feedback to communicate with the dog and the working model of the vest (Fig. 8.7). Further, it has specialized modes separately for different applications:

- Sports Mode: Optimized for high-activity periods, focusing on performance and exertion.
- Training Mode: Tailored for training sessions, monitoring progress and stress levels during drills.
- Mission Mode: Designed for operational use, prioritizing real-time data and discrete communication.

Fig. 8.6 App screens

Each mode adjusts the sensor sensitivity and data processing. For instance, Mission Mode emphasizes low-latency data transmission and haptic feedback, while Training Mode focuses on detailed behavioral analytics.

Some of the features incorporated in the vest:

- Haptic Feedback Commands: The incorporation of haptic feedback and a speaker into a dog vest revolutionizes the communication dynamic between a soldier and dog, fostering a more profound and efficient working relationship. Haptic feedback provides a silent, tactile language, allowing for discreet and immediate commands which are vital in sensitive operations where auditory signals are prohibited, cultivating a heightened sense of non-verbal understanding and responsiveness built on subtle physical cues.

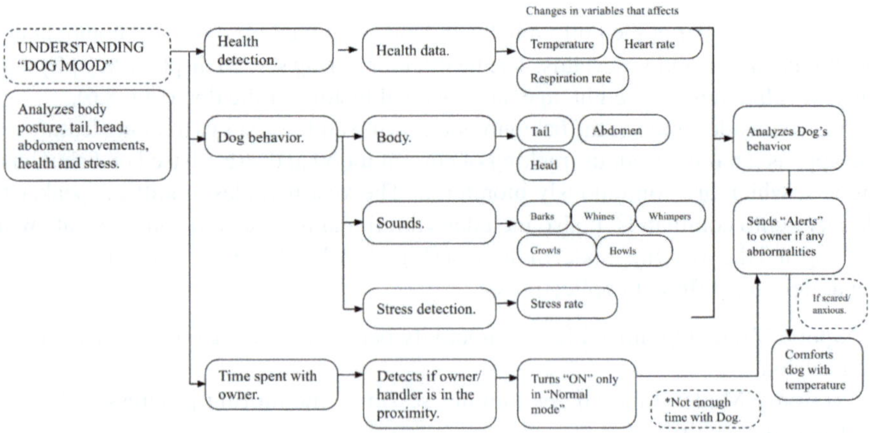

Fig. 8.7 Conceptual architecture of the vest

- Speaker for Commands: Remote verbal communication for guidance during missions, as the speaker system enables remote verbal communication, offering crucial guidance, reinforcement, reassurance from a distance, maintaining both the dog's focus and connection to the handler even when physically separated. Thus, it creates a multifaceted communication bridge that enhances trust, precision, and ultimately strengthens the unique bond.

The design is conceptualized in such a way that all the following factors were included.

- Biometric sensors which detect heart rate, respiration.
- Accelerometers and gyroscopes for activity and body language tracking.
- Thermal sensors for temperature regulation.
- A low-power microcontroller for data processing and communication.
- Bluetooth and cellular connectivity for real-time data transmission.
- Haptic actuators for vibration feedback, and a small speaker.
- A rechargeable, long-lasting battery.

This is a very well-integrated system, looking at the data that is processed and presented on a mobile app, translating it into actionable insights for the handler. It is specially designed considering the duties and responsibilities of a soldier. Initially, this involves understanding the duties of soldiers to comprehend the time constraints and requirements for them. A lot of thought has been put into understanding how to enhance their communication with minimal effort from handlers. The solution allows the handlers to track and monitor their dog's overall mood and behavior analysis, track the dog's vital information, receive real-time alerts and comprehensive reports, allowing them to make informed decisions about their dog's well-being and performance. Handlers or users can understand the dog's mood and behavior in real time, which helps them to respond and react quickly. This is crucial for understanding the dog's physical and mental changes over time, as ultimately, the proposed design concept can play a vital role in strengthening the bond between the handler and the dog, fostering a deeper connection built on trust and communication while being immersed in a broader ecology (Fig. 8.8).

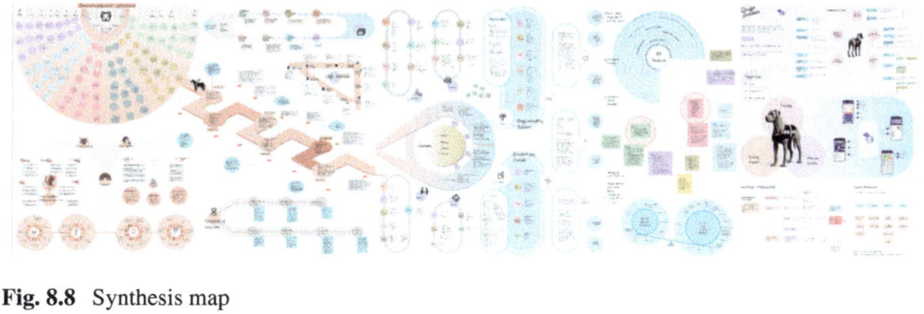

Fig. 8.8 Synthesis map

8.7 Conclusion

The journey began with a simple but powerful goal to strengthen the emotional and functional bond between service dogs and their owners. Through systems thinking and tools such as Gigamapping and ZIP Analysis, deep-rooted challenges such as stress, communication gaps, and behavioral issues that compromise these relationships were uncovered. By focusing on measurable components and stakeholder-specific pain points, a strategic leverage point was identified, i.e., real-time understanding of a dog's emotional and physical state. This led to the development of Chitta, a smart wearable vest integrated with a mobile app. Chitta tracks vital signs, emotional cues, behavior, and environmental conditions, offering tailored modes for training, missions, and daily routines. Its haptic and audio feedback features enable discrete communication, and its companion app delivers actionable insights to handlers, vets, and trainers.

The design is human and dog-centered, cost-effective, and built to improve not only individual dog-owner relationships but also to inspire systemic improvements in how working dogs are trained, monitored, and cared for. With the solution, the focus is not just on enhancing connections, but on redefining communication, care, and compassion in high-stakes environments. The design outcome is not just a product; it's a shift towards deeper understanding and higher standards in dog-human partnerships, and it is made possible by following systematic and systemic design processes.

References

1. World Animal Foundation (2026.) How Many Dogs Are In The World In 2026? https://worldanimalfoundation.org/dogs/how-many-dogs-are-in-the-world/. Last accessed 20 Mar 2026
2. Lea SE, Osthaus B (2018) In what sense are dogs special? Canine cognition in comparative context. Learn Behav 46(4):335–363
3. Haq AU, Malik HU, Randhawa SS, El-Shazly M, Chandran D (2024) Behavioural disorders in dogs and cats: a review. Agric Rev 45(3):514–519
4. India Today (2023) India's dogs of war: How Army recruits and trains its canine warriors. https://www.indiatoday.in/india/story/indian-army-dogs-kent-units-jammu-kashmir-terror-operation-2435877-2023-09-16. Last accessed 20 Mar 2026
5. Serpell J (ed) (2017) The domestic dog: its evolution, behavior and interactions with people, 2nd edn. Cambridge University Press, Cambridge
6. Hiby E (2013) Dog population management. In: Macpherson CNL, Meslin F-X, Wandeler AI (eds) Dogs, Zoonoses and public health. CABI, Wallingford, pp 177–204

Chapter 9
Supporting Research on Plastic Degrading Micro-organisms

Puspendu Adhikari, Ishita Srivastava, Rishabh Pandey, and B. M. Manoj

Abstract The primary aim of the project is to comprehend the possibilities of using microorganisms to aid in the degradation of plastics. The synthetic nature of the polymer structure of plastics causes it to degrade for centuries, leading to accumulation and long-term environmental damage due to the release of toxins. Microorganisms, such as fungi and bacteria, that are excessively exposed to plastics have now evolved to treat it as a source of nutrition, effectively digesting it into simple end products. During gigamapping, we first analysed the parts of the system, such as the stakeholders involved, impactful factors such as the life cycle of plastics, and interaction with microorganisms. Then we put them together to examine the system as a whole. Further, the data is analysed using ZIP analysis, where we zoomed into the highest impact areas, recognised pain points, and developed design interventions through innovations. In terms of the results, we have conceptualised a mycelium incubator and research kit specifically designed to study fungal behaviour on plastic degradation. This concept design solution unlocks a sustainable, scalable pathway for mitigating plastic pollution, bridging research gaps, and integrating bioremediation into the design by using the natural capabilities of mycelium to break down toxic materials. Our final solution included the design being more adaptable for independent researchers and allowing them to have different temperature and humidity levels to study the mycelium, to conduct more studies simultaneously, and generate more data.

Keywords Mycelium incubator · Microbial biodegradation · Circular economy · Sustainable waste management

P. Adhikari (✉) · I. Srivastava · R. Pandey · B. M. Manoj
Indian Institute of Technology Kanpur, Kanpur, India
e-mail: puspendua24@iitk.ac.in

V. Kant (ed.), *Human Centred Design for Systems*, Springer Series in Design and Innovation 68, https://doi.org/10.1007/978-3-032-30376-9_9

9.1 Introduction

Let's consider the following story about the vultures in India. Vultures are an integral part of this ecosystem and act as scavengers actively preying on dead and decaying matter. They help to control the spread of diseases through the dead matter. In India, there was a rapid decline in the population of the vultures during the mid-1950s as they were feeding on cattle that were treated with the anti-inflammatory drug known as Diclofenac. Unknown back then, this drug actually caused kidney failure in the vultures. This threw the entire ecosystem off balance as a rapid increase in diseases was witnessed, especially in areas where there was a huge number of livestock. As these livestock passed away, they also happened to pollute waterways that the whole population was taking drinking water from. As the scavengers in the system were removed, the rat and dog population increased as they fed on the dead carrion. This also led to an increased number of rabies cases. Not only that, but a socio-cultural impact on the population of the Parsi community living in India was noticed too. They are known to observe a funeral ritual known as Sky Burial, where they lay their dead on stone beds known as the Tower of Silence. In this ritual, they give up the dead back to nature by exposing them to the sun and to fauna, like vultures, to feed upon them. As there was a rapid decline in the population of vultures, this meant that the bodies were no longer decomposing as fast as they used to. This also forced the community to resort to other rituals like cremation or solar concentrators and give up their sky burial rituals [1].

This is a systemic problem with repercussions as the systemic challenges that are observed in various cases with the introduction of new technologies and solutions without assessing their impact can be observed. In this chapter, an interesting world of plastic-eating microbes was explored describing a systems design process of identification of key issues and possible solutions. The aim here is to understand how to use a human-centric systemic approach in solving a problem and come up with design solutions by understanding intervention points and leverage points that can be used to decode a problem and develop an optimal solution for a complex system. The key topic emphasized for this is the phenomenon of rapid evolution; concisely, the rapid evolution of microorganisms that have evolved to digest plastics.

This chapter begins with understanding case studies on rapid evolution, then exploring different aspects of plastics. This involves understanding their life cycle, their manufacturing processes, the varieties of plastics that exist, their impact on the environment and fauna, and lastly examining the microbes that digest plastics.

A technique known as gigamapping was used, where the parts of the system were addressed in detail followed by the system as a whole. Further, these insights were analyzed using ZIP analysis, where zooming into the highest impact areas, recognizing the pain points, and coming up with design interventions through innovations. In the end, the final design concept was discussed and how that intervention will impact the overall system. The process was divided into the following five major steps: framing the system, comprehending the system, envisioning interventions, formulating the interventions, and enabling the transition.

9.2 Framing the System

Our primary problem statement is to comprehend the plastic degradation microbes' ecology and formulate design interventions to leverage value for stakeholders. In order to dive into this problem, first, the need to recognize the revenge effects of plastics was addressed. "Revenge effects," a term coined by Edward Tenner, refers to unintended negative consequences that arise when a technology or system meant to solve a problem actually exacerbates it or creates new problems. This was also clear in the case of the vulture example [1]. While, in this case it refers to the impact plastics have on the ecosystem as they were introduced to it and their overall accumulation over time.

The term rapid evolution deals with any evolutionary changes in the flora or fauna caused by the presence or introduction of another entity in the system. These can be direct or indirect impacts on humans, such as climate change or plastic pollution. Here, microorganisms have been studied that have adapted to the new plastic-polluted environment and have evolved to digest their strong chemical bonds. Plastic-eating microorganisms are the result of rapid evolution in response to plastic pollution. Bacteria evolve quickly to adapt to their environment and to use available carbon sources, such as plastic. The term "rapid evolution" gained popularity in the late twentieth century, especially in ecological and evolutionary research on antibiotic resistance, climate change adaptation, and human-driven evolution [2]. Some "rapidly evolved" microorganisms, Ideonella Sakaiensis and Pestalotiopsis, digest plastics by developing special enzymes (e.g. cutinase) to break the complex polymer chains into smaller biodegradable compounds [3]. Several synthetic polymers have been validated to deteriorate within months through the action of a wide variety of fungal enzymes [4]. These enzymes can even help upcycle the end products into other products in the industry, thus converting the lifecycle of plastics into a cradle-to-cradle approach where all the stakeholders benefit economically and in terms of mediating adverse health effects [5].

Since numerous bacteria exist, they reproduce exponentially. Hence, the chances of random mutations and genetic variations also increase. Through this mutation and selection, bacteria can develop defensive mechanisms against antibiotics. They can alter the structures of the receptors so that antibiotics can't attach to the receptors in the host. One of the popular examples is MRSA, Methicillin-Resistant Staphylococcus Aureus, which is one of the most common antibiotic-resistant bacteria. These microorganisms that had evolved to have this ability were naturally selected and survived. Ideonella Sakaiensis was one of these bacteria that was found in a recycling facility in Japan. It was found in the sludge of this recycling facility, feeding specifically on PET plastics. In 2011, students at Yale University found out that this special kind of fungi that was usually found in the forest of Ecuador, known as Pestalotiopsis microspora, had now developed the ability to feed on plastics. In addition, they anaerobically digest these plastics even in landfills [6]. Based on these insights, the problem statement aims to find out the best leverage points through ZIP analysis in the cycle of plastics and the microbes that digest them. The approach comprised secondary research, synthesizing the information, and presenting it visually as a Gigamap.

In order to understand how plastics circulate within the economy and the ecosystems, there was a need to understand the life cycle of plastics and their interaction with the stakeholders. The life cycle of plastics (Fig. 9.1) was used as a guiding map to get an in-depth analysis of each stakeholder and intervention point. This study helped to understand the environmental impact and the direct and indirect exposure to toxins released at each step. The various pathways of plastics' disposal, along with the presence of microorganisms and their interaction with the plastics in their ecosystem was also mapped and observed.

The start of the cycle is the extraction of crude oil. This oil is formed by the organic matter that was long buried deep in the earth. This process takes place deep in the sea beds or the earth. It is very energy-intensive and requires a lot of heavy machinery and operations. After extraction, it goes through the process of fractional distillation, where it is broken down into different components like petroleum, naphtha, and gasoline. The naphtha is taken and is then further broken down into different hydrocarbon bonds or monomers such as ethylene, propylene, and styrene. These monomers are then put into a process known as polymerization, and here, these monomers will combine and form strong chemical bonds, which will then give the base of plastics.

The process of pellet production itself consists of many different steps. One of these steps starts with compounding and adding materials to stabilize it. It helps give

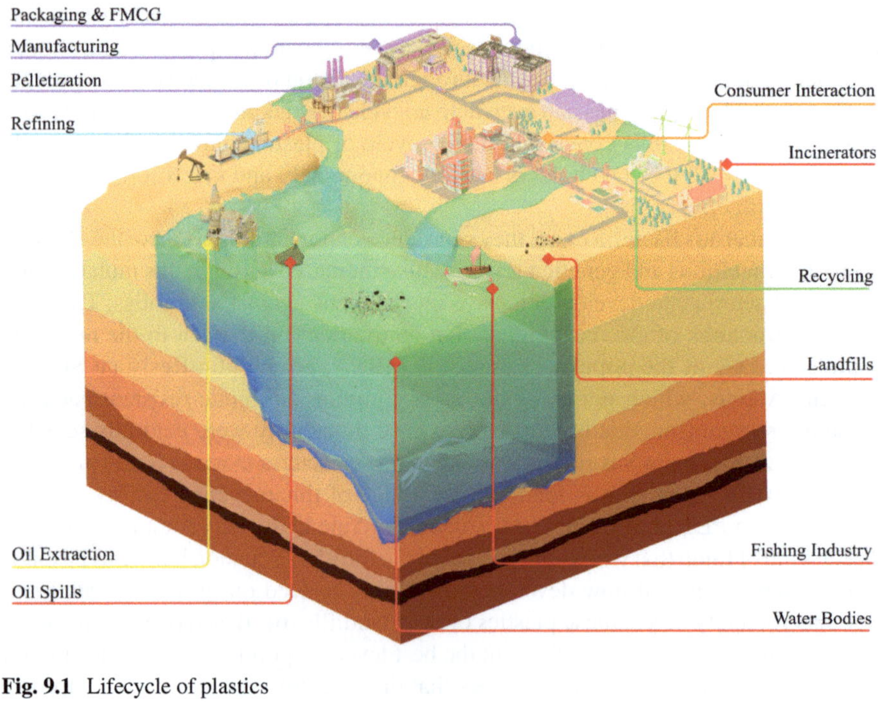

Fig. 9.1 Lifecycle of plastics

the compound a sturdy nature. Then, the resin is further melted and cut into small pellets. Excess moisture is removed, and they are shaped into a uniform size and quality.

These pellets are the starting unit of the plastic production cycle. These pellets are then shipped to different manufacturers, where they will be shaped into different objects like plastic bottles, furniture, or medical equipment. Once plastic is manufactured into different forms, it goes into the market for consumer use. Through the consumer and after industrial usage, it ends up in the ocean or landfills, where it is left untreated for a long time. Over the course of years, it breaks down into microplastics and enters the bodies of different fauna through various channels. A very small number of used plastics are actually being recycled for further reuse. This is where the packaging and fast-moving consumer goods industry becomes important. These plastic objects can be divided into different packaging materials, such as primary packaging, which will directly encase the product. Secondary packaging is used to pack more of the materials, such as shrink wrap, and tertiary packaging is used for bulk transport. More labels and packaging are added, and then they are shipped off to their end locations. After consumption, the packaging is discarded, of which a significant portion is transported to end locations using international waste disposal routes. These end locations such as the Southeast Asian and developing nations release the largest amounts of plastic from their coastal regions [7]. While these nations appear to be the primary polluters, they are not the primary plastic manufacturers. This disparity is explained by waste colonialism, where high-income nations export their plastic waste to developing countries, making the latter appear as the primary source of pollution [7].

9.3 Comprehending the System

The phrase "waste colonialism", refers to one set of people dominating another in their homeland using waste and pollution as strategic tools [8]. Developing countries may receive unusable and non-recyclable contaminated plastics, which can outweigh any financial benefits. For example, Malaysia receives plastic waste from wealthier nations, even though it produces much less plastic waste than these countries. In 2018, after China banned most waste imports, Southeast Asian countries such as Malaysia, Vietnam, and Indonesia saw a 500% increase in plastic waste imports from the U.S., U.K., and European Union [8].

Different kinds of plastics and toxins affect different parts of the human body such as the respiratory system through inhalation, on the skin through direct contact, and on the nervous system when consumed. In order to study the impact of toxins released by plastics, a comprehensive list of toxins released by plastics was made (Table 9.1). The list combined different types of toxins [9], how they are released, and what environmental and health impacts they tend to have. This helped to understand and identify the points of the highest impact and where the intervention can happen.

In the process of joining the charts together and making connections, the realization is that the plastics with the highest toxicity are the hardest to break down.

Furthermore, the environmental impact of plastics and their ultimate disposal was studied. Most plastics, around 50%, end up in landfills, and about 8–10% end up in the oceans. It goes on to show that the waste that is being picked up from the houses every day ends up being mismanaged and left out in open landfills, untreated. This waste can then pollute the soil and groundwater through leaching, and also biomagnify when consumed by fauna. As for the waste that ends up in the oceans, it is often noticed that it collects into patches. One of the examples is the Pacific Garbage Patch, a swirling mass of plastic debris spanning over 1.6 million square kilometers. And finally, only 9% ends up getting recycled [9].

Different properties of the types of plastics can be characterized such as durability, the time they take to degrade, waste volume, and their toxicity. Amongst these Low-Density Polyethylene (LDPE), comprises the largest amount of waste volume due to its massive use in the packaging industry (Table 9.1).

Further, based on summarizing secondary literature, the following insights are salient:

- Plastic waste is emitted to the Oceans.
- The maximum amount of plastic waste ended up in the landfills.
- Plastics cause pollution at every stage of their life.
- Packaging contributes the largest amount of plastic waste.

Plastic packaging has become an essential component of everyday life, integrated into almost every consumer product. The plastic pellets molded into packaging materials are produced through complex chemical processes derived from crude oil, specifically fractional distillation and polymerization processes. However, most plastics are not properly treated after use and end up in landfills or oceans, where they break down into microplastics, posing threats to wildlife and humans.

One way of addressing the situation is by using plastic degrading microbes. Most microorganisms choose a specific kind of plastic because these plastics have different polymer chains and need different enzymes to break them down [10]. They often use it as nutrition and release methane and water as the end products along with the monomers. Under aerobic conditions, microbial degradation turns plastic into CO_2, water, and microbial mass. This is one of the few processes where plastic is actually rendered into harmless materials, and that too by nature.

There are currently over 400 species of plastic-degrading microorganisms that have been identified. With the most notable examples being Ideonella Sakaiensis, which has now been found to degrade PET plastics, and Aspergillus tubingensis, a polyurethane-degrading mycelium. The plastic-degrading enzymes are mutations of natural ones. For example, PET-ase is a mutated form of Cutinase, which is used by microorganisms to break down cell walls. It is very crucial to study this cycle as it is the only naturally occurring way to break down the complex bonds of plastics. Human-made efforts have always resulted in toxic waste being generated as an end product. Whether it be incineration or recycling, they're never 100% effective [10].

Table 9.1 Toxins released by plastic [9]

Plastic type	Toxic chemicals	How toxins are released	Health and environmental impact	Common uses
PET (polyethylene Terephtha late)	Antimony, acetaldehyde	Heat exposure, recycling residues	Lung & heart toxicity, absorbs environmental toxins when reused	Water bottles, polyester fabric, food packaging
HDPE (high-density polyethylene)	Heavy metal stabilizers (older plastics)	Absorbs toxins from polluted environments, degrades into microplastics	Generally safer but contributes to plastic accumulation	Milkjugs, detergent bottles, pipes
PVC (polyvinyl chloride)	Dioxins, phthalates, vinyl chloride, Lead, cadmium	Manufacturing, heating, burning, ageing	Carcinogenic (dioxins, vinyl chloride), endocrine disruption, heavy metal contamination	Pipes, medical tubing, flooring, packaging
LDPE (low-density polyethylene)	Nonylphenols, hydrocarbons	Sunlight & wear release hydrocarbons, producing microplastics	Disrupts aquatic ecosystems, persistent plastic pollution	Plastic bags, shrink wrap, squeeze bottles
PP (polypropylene)	Plastic additives (antioxidants, stabilizers)	Degrades under UV light, absorbs pollutants	Low toxicity, contributes to microplastic pollution	Straws, food containers, automotive parts
PS (polystyrene)	Styrene, benzene	Leaches into hot liquids, degrades in sunlight	Neurotoxic, suspected carcinogen, major source of microplastic pollution	Disposable cups, food trays, insulation foam
PC (polycarbonate)	Bisphenol A (BPA)	Hot liquids, acidic foods, scratches & heating	Endocrine disruptor, linked to cancer, infertility, obesity	Reusable water bottles, medical devices
PFAS-coated plastics	Per- and Polyfluoroalkyl substances (PFAS)	Leaches into food & water, incineration releases toxic gases	Forever chemicals, causes cancer, immune suppression, metabolic disorders	Non-stick cookware, grease-resistant packaging

A list of the different kinds of bacteria and fungi, the enzymes they produce, the environments they are found in, and the plastics that they have specifically evolved to digest was noted (Table 9.2). Landfills and waste dumps tend to have a lot of these naturally occurring bacteria as they are also found in recycling facilities, as a lot of the plastic is melted into sludge, which is kept in chambers. A lot of soil and agricultural fields are littered with plastics, but the surrounding microorganisms have evolved to even use them as nutrients. Even in the areas of oil spills, these bacteria were found to have evolved to have the oil-degrading capacity for petrochemicals.

As these are found in various environments like landfills, rainforests, marine and ocean ecosystems, in varying capacities of oxygen availability, different pH systems, different surfaces, and different systems on which they work. Some of the fungi or mycelium that have developed enzymes and are found in different places like agricultural fields, decaying matter, compost, and marine surfaces, where they degrade different kinds of plastics were studied. After synthesizing all the secondary data, then mapping all the stakeholders involved in the cycle, categorising all the stakeholders as primary, secondary, and tertiary. All the stakeholders were visualized according to their impact on the system and their involvement (Fig. 9.2).

Then stakeholders were further visualized according to their knowledge of the system and the power they hold (Fig. 9.3). For example, a sanitation worker might have very low power and very low interest in solving this issue, even if he is impacted by it the most. Whereas the power of a documentary maker is very high, and it would also be in his interest to educate and create awareness.

Table 9.2 Different kind of plastic degrading bacteria [10]

Bacteria	Enzyme(s) produced	Found in	Plastics degraded
Ideonella sakaiensis	PETase, MHETase	Landfills, plastic waste sites	PET (polyethylene terephthalate)
Pseudomonas putida	Polyurethane hydrolase	Industrial waste, soil, water	Polyurethane (PU)
Bacillus cereus	Esterases, lipases	Contaminated soil, recycling plants	"PET, LDPE (low-density
Rhodococcus ruber	Alkane monooxygenase	Marine environments, soil	Polyethylene)"
Strepto myces spp.	Cutinases	Soil, compost, farmlands	Polyethylene (PE)
Lysinibacillus spp.	Proteases, Esterases	Agricultural fields, landfills	Polyester-based plastics
Alca nivorax spp.	Hydrocarbon-degrading enzymes	Marine plastic debris, oil spills	Polyethylene (PE), PET
Vibrio spp.	Polyethylene-degrading enzymes	Ocean water, plastic waste	HDPE (high-density polyethylene)
Geobacillus spp.	Polyesterases	Hot springs, compost	PE, PP (polypropylene)
Thermobifida fusca	Thermostable polyesterases	High-temperature waste sites	PLA (Polylactic acid)

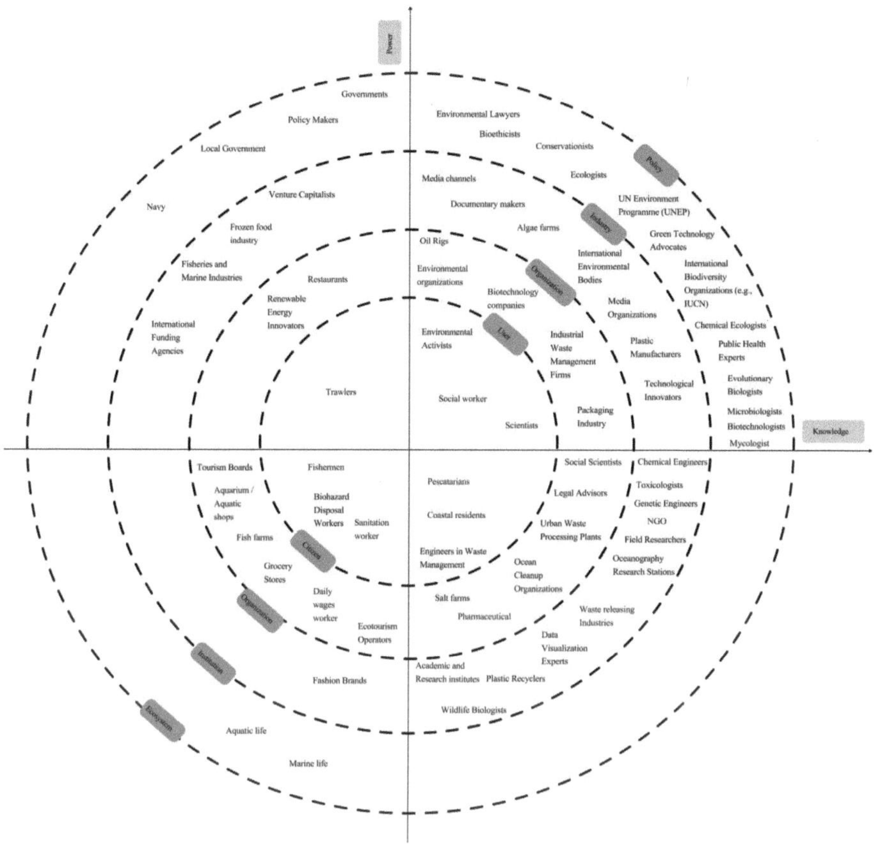

Fig. 9.2 Stakeholder mapping

Next, moving on to map the stakeholders into a more comprehensive map where they are categorized on different strata of individuality, organization, industry, and policy-making (Fig. 9.4). The quadrants are divided according to the power they hold and the knowledge they bear.

Further on, more dimensions and extremities are explored, mapping the characteristics of all the stakeholders involved. Here, the stakeholder bodies are not just noted down but also other factors like the plastic degradation capabilities of the microbes and the natural degradation time taken by plastics.

While these are not necessarily characteristics of the stakeholders, they are very important characteristics to note down. Marking down the extremities on them helps to categorize the most important, most impactful, and the least impacted entities within the system.

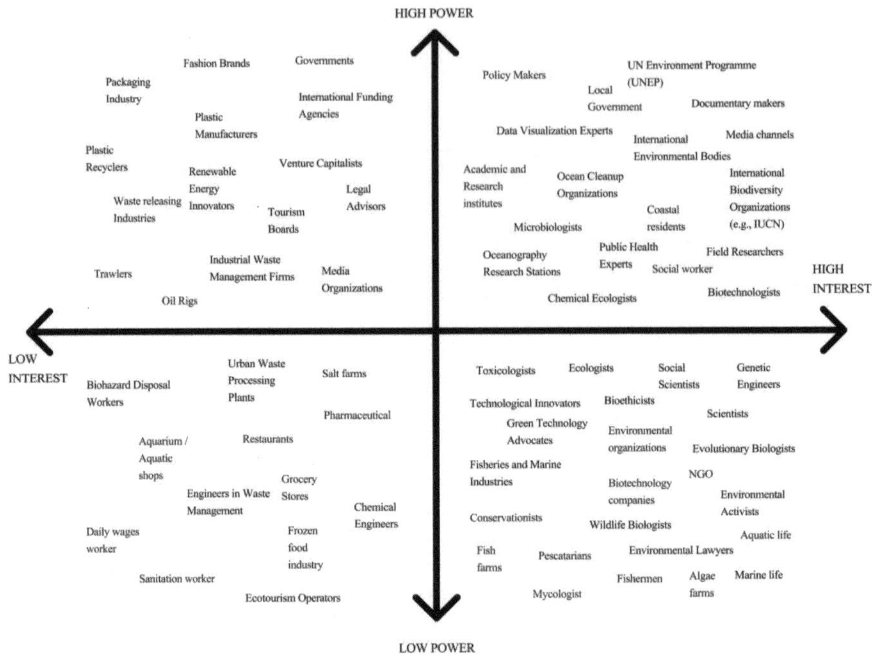

Fig. 9.3 Stakeholder mapping based on interest and power

9.4 Envisioning the Interventions

All the data collected through the background study was mapped and connected in the form of a Gigamap. This large-scale visual map allows to represent the complexity, relationships, and dynamics of the system in a tangible and analyzable format.

A ZIP analysis was conducted to delve deeper into specific subsystems and interactions within the broader ecosystem. Zooming into some of the aspects of the map and identifying several pain points where the system inefficiencies, user challenges, or environmental vulnerabilities were most evident. Some points that stood out the most were then marked as intervention points, representing critical opportunities for design innovation and strategic action. This evaluation helps to identify a leverage point where a small, focused intervention could lead to significant and cascading positive change throughout the entire ecosystem. Every prospect of controlling plastic pollution or reducing the impact on humans as well as the whole ecosystem was explored.

Intervention points were the ones that had the most impact and were the biggest deciding factors on where the plastics eventually end up. One of the intervention points was the plastic waste collection and segregation. Analyzing the scope for innovation for the collection and segregation of waste through a lot of diverse methods like auto-sorting bins, incentivizing the public to sort the plastic through collection booths and kiosks, and so on were discussed and analyzed further.

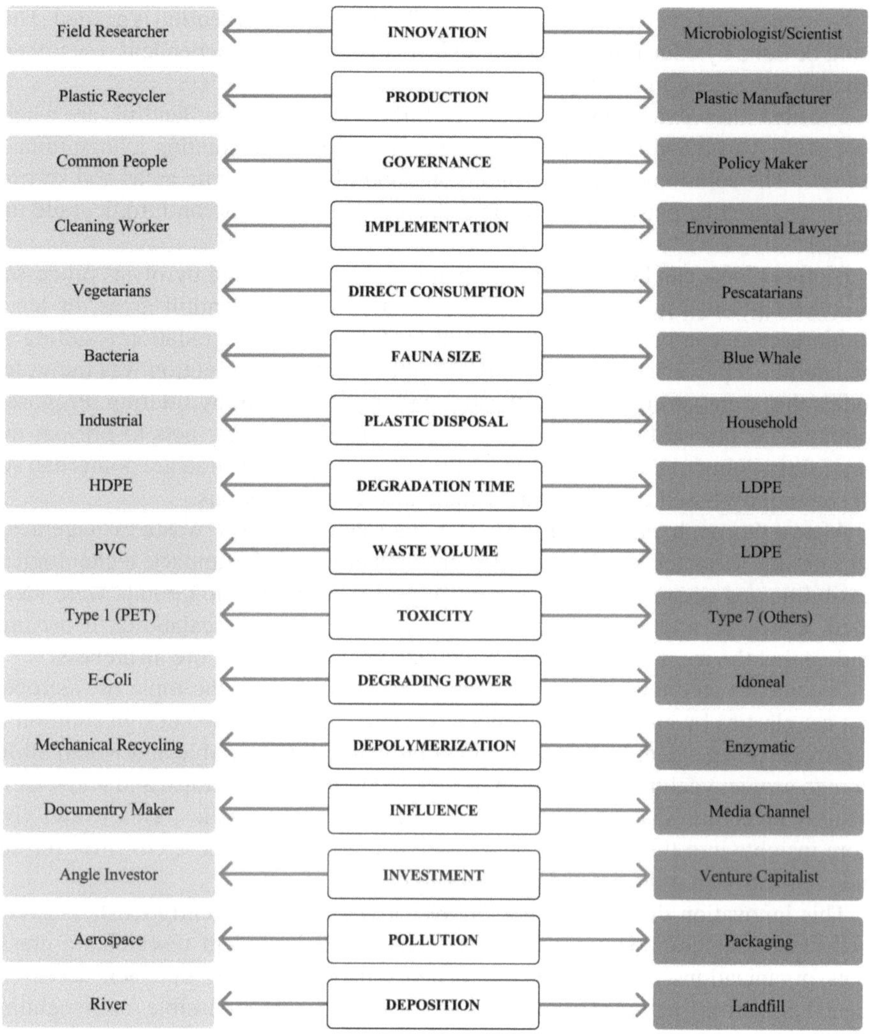

Fig. 9.4 Stakeholder variations based on characteristic

The next intervention point involved Waste Management at the Household level. A lot of solutions, and degradation devices at the household level were one among them. The next intervention point was developing and organizing the packaging of the product material. This could include bioplastics, which can be made from organic matter such as algae, bamboo, jute, coconut husk, amongst others. The plastics could also be manufactured with a self-degradation timer in them. When the consumer is done using them, they will put them in a certain situation, for example, in a chemical bath or in water, where they will degrade really quickly.

Awareness and education are one of the fields in which intervention is needed. This could be done by educating at the school level or by having independent researchers generate more data on this topic.

The next intervention point was the landfills. As noted earlier, landfills are where a lot of the waste ends up. Hence, utilizing this area meant creating a huge impact in itself. The solutions included zoning out landfills into specific areas and sorting out the waste, like plastics, and having a mycelium farm grow on it to degrade the plastics.

Biofuel plants can be set up using chemical degradation and pyrolysis processes to convert plastics to fuel at the nearby location of large landfill areas, or landfill bioreactors can be set up using chambers for bacterial degradation resulting in the landfills a power-generating point. Another point of intervention was the water bodies that were getting polluted through industrial waste or by littering. Proposed solutions include surface waste collecting robots and filtering nets to prevent the waste from going to the water bodies. Litter traps at different strategic points and set up community-based cleanup groups could also be the solutions.

After a thorough analysis of the system surrounding plastic waste management, considering variables such as manpower availability, social impact, technological feasibility, and economic viability, several potential intervention points were identified. However, after weighing their relative influence and scalability, it became evident that the most strategic leverage points lay in creating more awareness.

Specifically, there is a need for generating more data on the topic of microbes that eat plastics by enabling independent research. The proposed design solution is to provide independent researchers with a kit that will be portable and would allow them to grow mycelium in their own homes and in their own setups, and allow them to study mycelium that has developed enzymes to digest plastics. This would give more insights into the topic and have more data to magnify these enzymes further and come up with more natural solutions to degrade plastics.

This innovation directly bridges the research gap that currently exists between well-funded, centralized research institutions and independent researchers, grassroots organizations, and equally motivated citizen scientists, but lack access to specialized infrastructure. Through this approach, it becomes possible for mycelium-based research, particularly in areas such as plastic degradation and sustainable material development, to be conducted on a much broader scale, increasing the diversity and volume of data, insights, and local adaptations. This innovation would help in making the process of research decentralized, empowering independent researchers and citizens to contribute directly to plastic waste management. Consequently, it would help in creating a circular economy where plastic can move from a cradle-to-grave approach to a cradle-to-cradle approach. This is also low-cost and has a high impact, as affordability is a major consideration while designing the kits that will be circulated. This will contribute to bioremediation and using nature's resources to come up with a frugal solution.

9.5 Formulating the Interventions

Since plastic degradation was a key focus, the design of a portable mycelium growth kit for independent researchers was finalized upon. This design solution unlocks a sustainable, scalable pathway for mitigating plastic pollution, bridging research gaps, and integrating bioremediation into the design by using the natural capabilities of mycelium to break down toxic materials. This provided a direction to align ourselves with ecological intelligence by harnessing nature's own processes to co-create frugal, sustainable, and regenerative solutions, rather than imposing energy-intensive or technologically complex fixes. This intervention represents not just a tool, but a shift towards resilient, inclusive, and nature-inspired innovation.

In order to develop a viable kit, going on to the field and conducting primary research is important to gain deeper understanding. We interacted with a microbiologist who has been heading a bio-incubation lab for the past 14 years in order to gain the necessary details and knowledge. We learnt about the process of creating successful mycelium cultures. He also introduced the processes involved and gave various insights into what a researcher actually looks for when having the successful equipment to develop mycelium cultures. Some of the insights were that a laminar flow hood is one of the important things to be present in the lab, as it helps maintain a sterile environment and create a successful culture. This is vital for the preparation of culture media, inoculation of samples, and transfer of fungal spores or mycelial tissue, where even minor contaminants can outcompete or destroy the target culture. A good incubator with a shaking plate with a temperature range of 4–40 degrees is also important, as it allows one to study various conditions, as different fungal species require different optimal temperatures.

However, commercially available incubators with these features are typically designed for industrial or high-end academic laboratories, making them prohibitively expensive for small labs, as the entire incubator can only be set to one temperature and humidity setting. Hence, multiple experiments cannot take place inside the same incubator. An autoclave is needed to prepare the liquid broth medium that the mycelium will feed on. The broth acts as the primary growth medium for mycelium, ensuring it receives the necessary nutrients without competing with microbial interference.

The distribution and accessibility of fungal starter cultures also impact research scalability. These cultures can be provided in various formats, such as powder form, which is dehydrated spores of mycelium, live vials involving actively growing cultures, or dormant form, such as cold-stored or freeze-dried.

The role of designers also includes studying the limitations of existing incubator designs available in the market, apart from ideating for new incubator designs. The current incubators are pretty heavy and occupy a lot of space. Most lab incubators only regulate temperature, not humidity, leading to inconsistent or sub-optimal growth. Generic incubators may not offer adjustable ventilation or filtered air exchange, risking CO_2 buildup or contamination. Most incubators have fixed shelving and don't support custom setups, and don't have built-in IoT features, which are crucial for tracking temp/humidity/CO_2 fluctuations. Even maximum models don't support modularity or isolation between samples to prevent cross-contamination. So, the concept was to make the incubators smaller yet more functional.

For the ideation, starting with feature extraction from the insights and findings. The incubator should facilitate seamless interaction for lab technicians and researchers who often work long hours in high-precision environments. This included considerations such as intuitive placement of controls, comfortable viewing angles, smooth access to sample compartments, and minimal physical strain during operation. It should have advanced technology incorporated to enable precise control and documentation. The incubator design integrates smart sensors and automated data extraction tools, as these tools will continuously monitor vital parameters. The incubator should have a proper feedback mechanism through visual indicators and real-time alerts for monitoring the whole process of mycelium growth and culture.

During the ideation phase of the product development process, several innovative design concepts were tailored specifically for efficient mycelium cultivation and monitoring. One of the key proposals that emerged was a cylindrical incubator design, chosen for its compact form factor, spatial efficiency, and functionality. Here, all the samples can be easily accessed as they are stacked on top of each other. This will make it suitable for both laboratory settings and small-scale production environments. Incorporated digital interfaces and smart monitoring features into the concept to enhance functionality and align with contemporary technological trends. The display and the control panel are positioned on top for better accessibility. A transparent fiber-reinforced body was also proposed, allowing users to visually monitor the internal environment without compromising the containment or hygiene. The clear material choice supports passive observation and complements an aesthetic appeal suitable for modern lab environments.

Recognizing the dynamic nature of mycelium growth, which can span from days to several weeks, a set of IoT-enabled features aimed at automating data collection and remote observation. These include integrated cloud storage for data logging, environmental condition tracking, and real-time alerts in case of deviations from preset parameters. These features enable remote monitoring via smartphones or computers, reducing the need for physical intervention and enabling consistent control of mycelium culture environmental parameters.

To push the boundaries of non-invasive observation further, a proposal to integrate a small, high-resolution microscope coupled with an external digital camera system where users can document growth patterns, perform time-lapse analysis, and share visual data with collaborators or research teams seamlessly was given priority because of its added advantage. These ideation elements reflect a strong emphasis on user-centered design, merging functional efficiency, compact aesthetics, and technological augmentation to support a smarter innovation.

This kit (Table 9.3) will contain all the basic necessities needed to conduct research on mycelium and grow it at home. It will have all the tools such as petri dishes, scalpels, inoculation loops, pipettes, beakers, and vials. It will also be provided with all the chemicals needed, such as the starter for the mycelium, the feeding broth, and the plastic substrate. It will also have equipment such as the incubator, a UV laminar flow, a collapsible still air chamber, a tiny microscope, and even the possibility of a plastic shredder for any experiments on the kinds of plastic they own. An additional enzyme study kit can be provided to the researcher if they want to dive deeper into the enzyme study. This would include the pH meter, the Spectro color meters, the lysis buffer, and even fluorescent dyes.

The final solution is named Hyphae Hatch (Fig. 9.5) as it's a design that is made more adaptable for an independent researcher and allows them to have different temperature and humidity levels to study the mycelium. The incubator will have different chambers, which could have different temperature and humidity settings, and the incubator would accommodate all of them, as researchers often work with multiple strains in parallel. Hence, one of the units can have different conditions that could be studied at the same time. This would help the independent researcher conduct more studies at the same time and generate more data.

Hyphae Hatch is a portable, affordable research kit that lets individuals grow and study plastic-eating fungi. The kit democratizes science, bridges research gaps, and empowers citizen scientists to drive innovation in bioremediation, as it is necessary to develop an effective plastic biodegradation process to accelerate the natural degradation rate of plastics [10, 11]. It will take up less space, have the accommodation for a shaking plate, and have an external camera to monitor the mycelium growth. Every chamber has a separate digital display, and all the controls will be performed from the outside. One of these chambers would also be designed to support high-heat dry sterilization cycles in order to have a replacement for an autoclave to prepare the feed broth. It has a transparent glass front door for easy monitoring of broader aspects from the outside. Its metallic body makes it sturdy and stable while also being in a shaker mode (Fig. 9.5).

While the kit is also provided with a modular laminar hood to remove the petri dishes and inoculate them in a sterile environment. The kit will also ensure safety and include safety instructions and a manual for novices and first-time researchers. The

Table 9.3 Hyphae hatch-kit requirements

Tools	Culture components	Equipment	Safety	Miscellaneous
Plates+ petri dishes	Pre-prepared nutrient agar	Portable incubator	Gloves	Notebook
Syringe	Liquid feed solution	HEPA filter/laminar flow	Masks	Instruction manual
Scalpel	Starter	Plastic shredders/ pellet maker	Ethanol spray	Markers
Inoculation loop	Plastic pellets		Butane flame torch	
Ziploc bags			Hydrogen peroxide	
Glass jar			Collapsible still air chamber	
Vials				

Fig. 9.5 Hyphae hatch—3D renders

entire kit, along with the incubator, will be packaged in a bio-safe sturdy packaging, which would ensure safe transport and also portable storage.

9.6 Enabling the Interventions

Zooming out a bit and asking, how is this project contributing to addressing the plastic pollution problem on a larger scale? To understand this, a new stakeholder map to assess the impact of the intervention was visualized. In this map, the researchers are in the middle of it and hold the most power. This change in power may also provide an impetus to the waste management industry as various researchers can come up with innovative and efficient concepts. A potential solution could be enzymatic degradation of the plastics that was successfully developed, then the waste management industry will be pressured to deploy it [12, 13]. The government and policymakers will also have to put in effort, along with the researchers, to push through these solutions. Solutions such as these will also have a huge impact on the flora and fauna, especially it will have a huge impact on humans who are currently suffering from various diseases or are impacted daily by the consequences of being in touch with plastics.

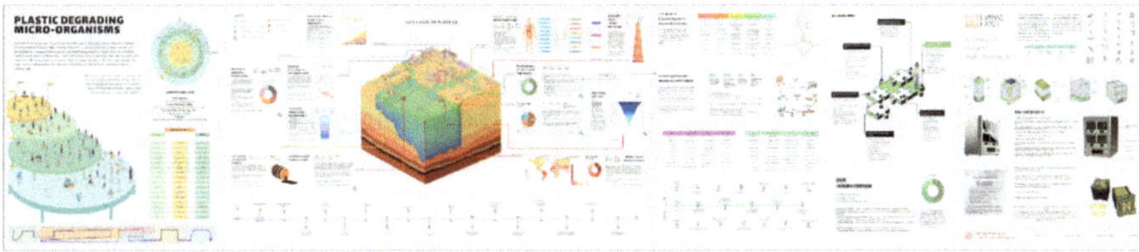

Fig. 9.6 Synthesis map

With the intervention, the researcher will have sufficient data to intervene in a more application-based solution to fight plastic pollution using plastic-degrading microbes at the following stages

- Plastic-eating mycelium farms at landfills
- Degradation device for households
- Mycelium compost for plastics at home
- Bioplastics for packaging
- Self-degrading plastics

A synthesis map takes one through the story of the lifecycle of plastics through visuals mapped at each process, accompanied by infographics and information at every stage (Fig. 9.6). First, looking into the stakeholders and their characteristics, they are represented in different tiers and distributed into quadrants according to their knowledge and power. Next, seeing all the stages and also the presence of plastic-degrading microorganisms at each stage. This helps us to understand how each process contributes to the plastic pollution problem and the extent of it. Further, we dove deeper into the plastic-degrading microbes and the environment in which they are found. A timeline visualization helps to see the stark difference between the environments and degradation time. Next, the different intervention points were visualized, with several iterations of ideation. The features were finalized keeping in view all the key considerations for the incubator, which along with the researcher kit was the final solution.

9.7 Conclusion

Using a systems design approach, a key leverage point was identified: enabling independent research on plastic-degrading fungi. This solution was called Hyphae Hatch; a portable, affordable research kit that lets individuals grow and study plastic-eating fungi. It is equipped with essential tools, an adjustable incubator, and safety features. The kit democratizes science, bridges research gaps, and empowers citizen scientists to drive innovation in bioremediation. With the proposed solution, plastic waste is turned into an opportunity, pushing towards a circular, sustainable future, one home lab at a time. Plastic pollution is a global crisis, but what if the solution starts at home? It is a question left for everyone to think about. The important takeaway is that a human-centred system design approach enabled to systematically scope the problem and then provide a solution. Just like the vultures, the story of the plastic-eating microbes becomes more tractable if approached from a systemic viewpoint rather than a symptomatic viewpoint.

References

1. Frank E, Sudarshan A (2024) The social costs of keystone species collapse: evidence from the decline of vultures in India. Am Econ Rev 114(10):3007–3040. American Economic Association
2. Ellner SP (2009) Lectures on theoretical ecology. Cornell University, New York
3. Russell JR, Huang J, Anand P, Kucera K, Sandoval AG, Dantzler KW, Strobel SA (2011) Biodegradation of polyester polyurethane by endophytic fungi. Appl Environ Microbiol 77(17):6076–6084
4. Khatua S, Simal-Gandara J, Acharya K (2024) Myco-remediation of plastic pollution: current knowledge and future prospects. Biodegradation 35(3):249–227
5. Tamoor M, Samak NA, Yang M, Xing J (2022) The cradle-to-cradle life cycle assessment of polyethylene terephthalate: environmental perspective. Molecules 27(5):1599
6. Russell JR, Huang J, Anand P, Kucera K, Sandoval AG, Dantzler KW, Hickman D, Jee J, Kimovec FM, Koppstein D, Marks DH, Mittermiller PA, Núñez SJ, Santiago M, Townes MA, Vishnevetsky M, Williams NE, Vargas MP, Boulanger LA, Bascom-Slack C, Strobel SA (2011) Biodegradation of polyester polyurethane by endophytic fungi. Appl Environ Microbiol 77(17):6076–6084. American Society for Microbiology
7. Meijer LJJ, van Emmerik T, Van Der Ent R, Schmidt C, Lebreton L (2021) More than 1000 rivers account for 80% of global riverine plastic emissions into the ocean. Sci Adv 7(18)
8. Stoett P (2024) Plastic waste colonialism: a typology of global toxicity. In: Plastic waste trade: a new colonialist means of pollution transfer. Springer Nature Switzerland, Cham, pp 3–15
9. Lithner D, Larsson Å, Dave G (2011) Environmental and health hazard ranking and assessment of plastic polymers based on chemical composition. Sci Total Environ 409:3309–3324. https://doi.org/10.1016/j.scitotenv.2011.04.038
10. Ekanayaka AH, Tibpromma S, Dai D, Xu R, Suwannarach N, Stephenson SL, Karunarathna SC (2022) A review of the fungi that degrade plastic. J Fungi 8(8):772
11. Bora DK (2020) Rise of the sustainable circular economy platform from waste plastics: a biotechnological perspective. MRS Energy Sustain 7(1):E28
12. Srikanth M, Sandeep TSRS, Sucharitha K, Godi S (2022) Biodegradation of plastic polymers by fungi: a brief review. Bioresour Bioprocess 9:42. https://doi.org/10.1186/s40643-022-00532-4
13. Temporiti MEE, Nicola L, Nielsen E, Tosi S (2022) Fungal enzymes involved in plastics biodegradation. Microorganisms 10(6):1180

Chapter 10
Participatory Design and Social Innovation for Permaculture

Agnivesh Sharma

Abstract Permaculture is considered a sustainable alternative to conventional industrial agriculture, but it is still not widely practiced among the Indian marginal farmers due to a lack of knowledge, comprehensibility, and the lack of widespread training. This project demonstrates the co-design of a participatory permaculture toolkit that will help farmers to make proper and effective decisions when it comes to planning and implementing sustainable agricultural practices. This toolkit was developed iteratively over two years and uses visual aids, interactive models, and farm mapping techniques to support learning. The design process involved multiple co-creation workshops with farmers and designers, refining the toolkit with real-world feedback. The toolkit has multiple elements that can be used for farming. These include various trees, crops, farm elements such as water source, amongst others. These elements ensure that realism of the real-world settings are retained, these elements can support farmers cognitively, as well as farmers can relate to the toolkit while using it. This research work describes the design process, the design iterations, and the final design that can be used as a reference model by policymakers, researchers, and agricultural practitioners by involving co-design to support sustainable farming transitions in India.

Keywords Permaculture · Co-creation workshop · Participatory design

10.1 Introduction

Imagine a city that wants to design a new public park. A traditional design approach might have city planners and architects draw up a plan, then reveal it to the public when the park is already being built. In contrast, in participatory design, involving local residents from the start, asking families what they would want in a park, maybe holding community workshops where neighbors brainstorm ideas with the designers.

A. Sharma (✉)
Indian Institute of Technology Bombay, Mumbai, India
e-mail: agnivesh@iitb.ac.in

V. Kant (ed.), *Human Centred Design for Systems*, Springer Series in Design and Innovation 68, https://doi.org/10.1007/978-3-032-30376-9_10

Parents might say they need a safe playground, the elderly might want plenty of benches and shaded areas, and teenagers might request a basketball court.

This way, the design process will be inclusive, and the design will be created based on a comprehensive understanding. The role of the designer is emphasized as moving beyond that of a creator to a co-creator; participatory design embodies that viewpoint.

Designers have increased their focus on sustainability and a better future for the upcoming generations [1]. Here, in this chapter, we demonstrate how participatory design principles and techniques can be used systematically in sectors such as agriculture to bring about sustainable solutions with elaborate details on how designing with people, not just for them, can lead to better, more meaningful solutions in social systems. The idea "Nothing about us, without us" really captures this spirit; it is to emphasize: that we cannot design solutions about people's lives without including those people in the conversation.

In simple terms, participatory design means including the end-users or community in the design process from the very beginning. Instead of designers working in isolation and then handing a finished product to people, participatory design invites those people who will use or be affected by the product/system to collaborate during design. This chapter describes the importance of participatory design where the analysis and interventions are focused on a social system of agriculture. This chapter demonstrates how a Participatory Permaculture Toolkit is co-designed to help marginal Indian farmers practice sustainable farming that has been co-designed. This toolkit was created through a collaborative design process with the farmers themselves.

Participatory design functions as a democratic practice that produces solutions that are grounded in the reality of the community (Fig. 10.1). This involves not only hearing people's thoughts through interviews, which they can easily express, but also when watching how they work or interact, uncovering what they might not say out loud. Participatory design goes beyond the surface to truly understand people by involving them in the process of making and reflecting.

Participatory approaches started in the 1970s in Scandinavia, where designers collaborated with workers and citizens to develop systems and environments. The purpose was to create results that both meet user requirements and give users power. It's not just about what people say, but also what they do, and what they deeply feel or imagine, their tacit needs, dreams, and desires. In addition, by co-creating with them through sketches, models, or role-play, tapping into deeper insights into their unspoken needs, dreams, and intuitive knowledge.

10.2 Framing the System

The key principles of participatory design involve the core idea of collaboration, where designers and users are partners. This requires mutual respect, as the users are seen as experts in their own needs and experiences, while designers bring their design skills to facilitate and create. Another principle is transparency; everyone should

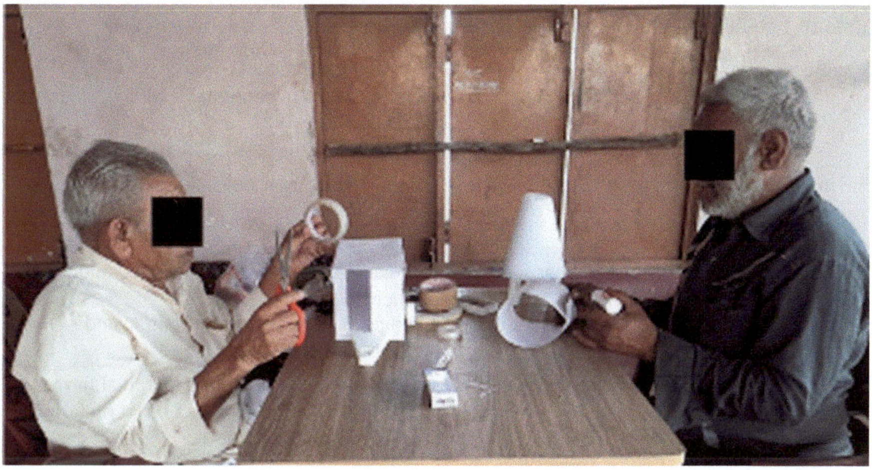

Fig. 10.1 Engaging farmers in design process

understand the process and have a voice. There is also an emphasis on iteration, where ideas are tried out (through simple prototypes or sketches), feedback is gathered from participants, and things are refined together.

So, how can incorporation of participatory design for agriculture be done? Agriculture is the backbone of human sustenance and civilization, but "Where is the agriculture that can feed the hungry billions of the world?". Large-scale mechanization [2], excessive use of insecticides and pesticides, and the widespread practice of monoculture have created a huge negative impact on the natural cycle and farmers' lives.

Soil contain millions of beneficial microorganisms that are essential for healthy plant growth. But with the continued use of chemical inputs, unintentionally destroying this rich biodiversity beneath the feet. The result? Weakened soil, declining yields, increased costs, and a cycle that is neither sustainable for the environment nor for the people working in the land. There are many other problems related to the current chemical-based agriculture, like ill-health, soil degradation, malnutrition, unemployment, migration farmers' suicides [3].

In India, farming is dominated by smallholders. In fact, over 80% of Indian farmers operate small holdings, meaning they have only a few acres of land or even less. A "marginal farmer" typically has very limited land and resources, often farming just enough to feed their family and maybe sell a small surplus (Table 10.1). These farmers often have traditional knowledge and a close relationship with their land. They usually have low incomes, and many have limited formal education or access to up-to-date agricultural information. Because of these constraints, marginal farmers are highly vulnerable, as a bad harvest or a sudden expense can be devastating. Climate change (like unpredictable monsoons or droughts) and market price swings hit them particularly hard.

Table 10.1 Indian agriculture marginal farmers' challenges

Socio-economic challenges	• Poverty and indebtedness • Lack of access to institutional credit • Land ownership issues • Landless laborers
Environmental and climatic challenges	• Impact of erratic monsoons, floods, and droughts • Soil erosion, depletion of groundwater, and declining soil fertility
Technological and infrastructural barriers	• Limited access to modern farming techniques and equipment • Challenges in storage, transportation, and marketing of produce
Policy and institutional challenges	• Government policies affecting marginal farmers • The role of middlemen and its impact on farmer profits

The big question is, how to help them improve their livelihoods? This is where sustainable farming practices (such as permaculture) could be beneficial, but adoption isn't straightforward [4]. They also face many challenges, involving Socio-Economic Challenges, Environmental and Climatic Challenges, Technological and Infrastructural Barriers, Policy and Institutional Challenges.

There is a clear need for sustainable farming approaches and transition to sustainable practices, specifically to restore soil health, such as composting, mulching, and organic methods that improve fertility. Also, to conserve water, rainwater harvesting, ponds, contour farming, and mitigation of droughts is required. To enhance biodiversity, one has mixed cropping and agroforestry, which boosts resilience to pests/weather. Building climate resilience and reducing dependence on costly external chemical inputs.

Examining sustainable agriculture itself is very broad, involving many aspects. It is not just about growing crops. It's about balancing three key dimensions: economic, social, and environmental. Firstly, an economy where farming isn't profitable won't last. Farmers need to earn a decent income to keep going. Second, a social environment where the well-being of farmers, their families, and rural communities matter. Sustainable farming practices support their health, dignity, and ways of life. And third, environmental, where the need to care for the land, water, and biodiversity. If the resources get depleted, safety of the future generations is compromised. True sustainability in agriculture only happens when all three pillars work together.

10.3 Comprehending the System

Agroecology (Fig. 10.2) starts with diversity, not just of crops, but of knowledge, people, and practices. When co-creation and sharing of knowledge happens, the farmers, scientists, and communities learn from one another. Synergies and efficiency help to make the most of using natural interactions and smart resources instead of depending on chemicals. Then comes recycling, turning farm waste into something valuable, like compost or mulch, instead of letting it go to waste. Resilience is about being able to bounce back from droughts, pests, or market shocks, and that ties closely to social values, putting people, equity, and community at the heart of farming. Culture and food traditions reminded that farming isn't just a job, it's a way of life, deeply tied to heritage. For this system to work, the need for responsible governance where farmers are involved in decisions and a circular economy that keeps things local, reduces waste, and supports community livelihoods. Together, these elements show that farming isn't just about food, it's about people, the planet, and a shared future.

One might wonder, if sustainable farming is so great, why isn't everyone doing it? There are several barriers to adoption, especially for marginal farmers. Firstly, sustainable farming is knowledge-intensive and complex. It's not as straightforward as following a prescribed formula, like "apply X kg of fertilizer, plant Y seeds". Instead, it requires applying a set of principles, observing one's land carefully, and designing a custom system. For many farmers, especially those with limited formal education or exposure, this learning curve is intimidating. Secondly, there is a lack of structured training available. So, most small farmers haven't even heard of sustainable farming or seen it in practice. Third, adopting sustainable farming involves initial effort and experimentation as one has to invest time in redesigning the farm layout, perhaps planting trees or new crop combinations that might take a while to pay

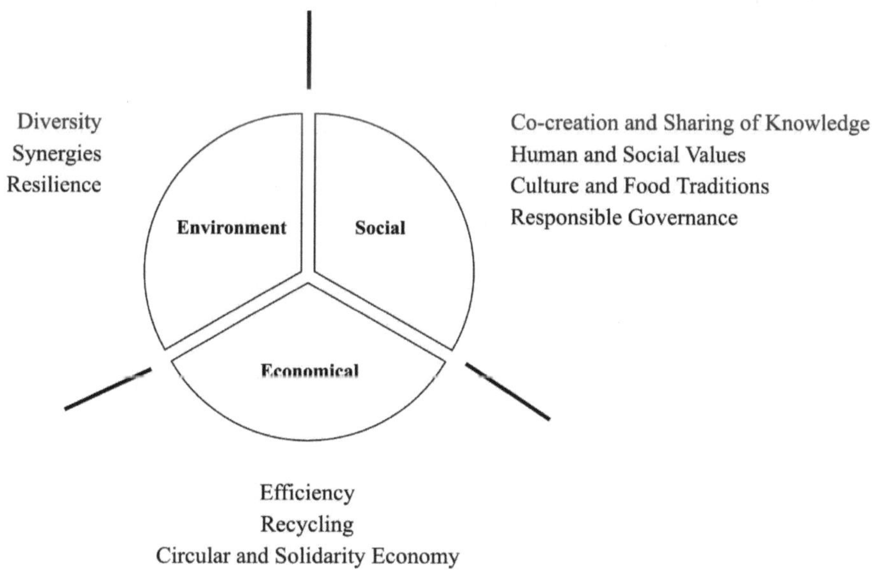

Fig. 10.2 Agroecology elements—social, economic, and environmental aspects

off. There's also the issue of perception and trust, as many farmers are understandably skeptical of new or different methods. They might view sustainable farming or organic techniques as "unproven" or less productive. It can be hard to break habits, and there may not be local examples to follow. Also, resource constraints are always a barrier, as a marginal farmer doesn't have extra land to experiment on or spare money to invest in new tools or seeds that might be needed during the transition.

In this project, one piece of this puzzle as the knowledge and training gap, by using participatory design of a permaculture toolkit was addressed. But what is permaculture? The term comes from "permanent agriculture," and later was expanded and broadened as an inclusive term to "permanent culture". This movement was started by Bill Mollison and David Holmgren in the 1970s [5]. Permaculture represents a sustainable agricultural and community design method that creates farming systems that replicate natural ecosystems. A permaculture farm operates with natural processes while avoiding monoculture systems and heavy external chemical inputs. The philosophy follows a holistic approach by linking land with water, plants, animals, and people to create mutual support systems where compost from one element serves as a resource for another. The goal is a self-sustaining, regenerative system, as permaculture is guided by a set of ethics and principles that shape every design decision. At its core are three key ethics involving Earth Care, People Care, and Fair Share. While permaculture offers a sustainable alternative to industrial agriculture, its adoption remains limited among Indian farmers, particularly small and marginal farmers [6].

The complexity of permaculture design, the knowledge-intensive nature of its principles, and the absence of structured training programs pose challenges to its widespread implementation [5]. The adoption of permaculture by small and marginal farmers is usually limited by complexity, lack of access to systematic instruction, and infrequent exposure to organized training workshops [7]. However, while permaculture principles are based on the idea of non-dependent, maintenance-free agricultural systems, many farmers fail to understand how to apply these principles to the specific situation of their land [4].

Earth Care means prioritizing the health of the environment, protecting the soil, conserving water, and supporting thriving ecosystems. People Care focuses on the well-being of individuals and communities, ensuring that human needs are met in a respectful and supportive way. Fair share, also known as "return of surplus," emphasizes the importance of sharing abundance and setting limits on consumption, so resources can be distributed equitably and sustainably. These ethics are the foundation of permaculture thinking and practice.

There are 12 design principles, developed by David Holmgren and Bill Mollison [8]. While not all the twelve were considered here, a few key examples include:

- Observe and interact, learn from nature before taking action.
- Catch and store energy, harvest resources like rainwater and sunlight for future use.
- Produce no waste, design systems where everything is reused, and nothing goes to waste.

- Use and value diversity, integrating different plants, crops, and species for a more balanced, resilient system.

These principles serve as tools for thinking as they help in designing agricultural systems that are productive, resilient, and adaptable to change. Together, the ethics and principles of permaculture promote sustainability not just in environmental terms, but also socially and economically, ensuring long-term well-being for both people and the planet. In addition, extensive literature studies were done, trying to identify some of the sustainable farming practices in India and their relevance to the 10 elements of agroecology.

As designers, to truly design for farmers, the need to understand permaculture from within, not just from books, and enrolling in a Permaculture Design Certificate (PDC) course at a Permaculture Farm, one of the pioneers of the movement in India, added to the overall understanding. Over the course of 13 immersive days, one can get hands-on with everything from soil restoration and composting to water harvesting, farm planning, and design mapping (Fig. 10.3).

From this course, it was quite clear that it wasn't just about learning techniques; it was about experiencing the ethics, observation-based thinking, and systems approach that permaculture stands for. The knowledge from this course helped to ground the research in real Indian farm conditions and made sure it's not just applying theory but co-creating with deep, context-based insight.

Permaculture can be divided into 5 steps. The first step begins with sector and site analysis to understand how sunlight, wind, water, and slope affect farm layout. Then comes needs and yields analysis, where mapping out what each element in the system gives and takes, whether it's plants, animals, or people. Water harvesting is a major focus of designing ways to spread and store water naturally. The last step includes zone planning, dividing the land into zones based on how often different

Fig. 10.3 Permaculture design course

areas are used, so that energy and resources can be managed efficiently. After knowing the conceptual farm plan, translating permaculture theory into action in an Indian context was a new experience. The permaculture training could not be based on a single model as India has different agro-climatic zones [9].

One of the most powerful takeaways from the permaculture course was understanding the difference between Principles and Ethics. The ethics and principles of permaculture, like Earth Care, People Care, and Fair Share, are universal. They stay the same, no matter where one is in the world, but the planning and strategies, those that change from farm to farm. What works in a dry region won't work the same way in a hilly or flood-prone area. So, while every design involves steps such as sector analysis, water harvesting, and zone planning, how one applies them depends on the local context, the needs of the farmer, and the specific landscape. This balance between consistent values and flexible solutions is what makes permaculture so powerful and so adaptable for Indian farming systems. To truly understand how permaculture works on the ground, over a year was spent working closely with a farm near Ujjain (MP), around 2.6 acres in size. It was more than just a field study; it was an intrinsic longitudinal case study, where the key observations were about how the farmer was transitioning from conventional to permaculture practices over time.

This entire case study was based on two farming cycles, Rabi and Kharif, where wheat and soybean crops were observed. This study helped to understand farmers closely and their needs and knowledge. This immersive experience gave deep insights into both the technical challenges and the mindset shift needed for a farmer to adopt permaculture, as these insights helped in directly shaping the toolkit design.

Ethics & Principles ——————→Remains same.

Planning & Strategies ——————→Changes from farm to farm.

As part of the longitudinal case study, the farmer's entire journey across two crop cycles, Rabi (wheat) and Kharif (soya-bean), to understand what permaculture looks like throughout the year was documented (Fig. 10.4).

10.4 Envisioning the Interventions

Valuable insights as to why permaculture adoption is still a challenge for many Indian farmers were uncovered. Firstly, permaculture is knowledge intensive. It's not just about planting; it draws from ecology, soil science, botany, and more, resulting in challenges without support. Second, training is essential. Farmers need to learn not just the principles, but also how to design, implement, and evaluate a whole system which is tough if they haven't had prior exposure. There is also a major gap that is evident in the form of community support.

In many villages, the social systems such as local food networks that make permaculture work are missing. Without this, farms struggle to sustain themselves. Another challenge is policy, where a lack of targeted subsidies or incentives for farmers who want to shift toward sustainable methods like permaculture can create a risk in the early stages. Farmers having a lot of embedded wisdom or tacit knowledge, and

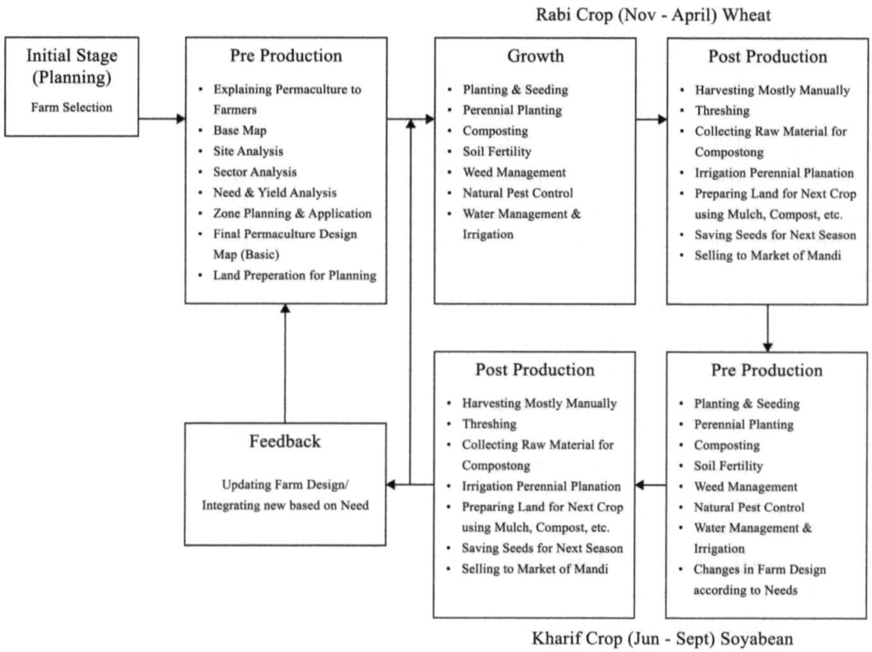

Fig. 10.4 Map of various activities at different stages in permaculture farming [6]

hands-on experience with the land means could pick up permaculture quickly, if the right support systems were designed. This understanding directly shaped how the toolkit was conceptualized, which was: a bridge between existing farmer knowledge and the structured systems thinking of permaculture. Given these challenges, participatory design, co-designing with the farmers themselves, was considered as a key to bridging the gap.

Co-design tends to produce solutions that are relevant and user-friendly. As farmers possess deep local knowledge, they understand their soil, climate, crops, and community far better than any outsider. Any solution for sustainable farming has to fit those local ground realities by involving farmers in the design of the toolkit, tapping into that knowledge. When the eventual users, in this case farmers, help create a tool, they naturally shape it to suit their preferences and constraints. This means the toolkit would use language, visuals, and processes that make sense to them, as opposed to something designed in a studio that might be too abstract. Also, involvement brings a sense of ownership. A farmer who helped design a toolkit is much more likely to trust it and use it, because it's their creation too. It's not a top-down directive; it's something they had a hand in, which in turn can greatly improve adoption. Co-design also provides a platform to merge scientific knowledge with traditional wisdom. Here, designers bring permaculture concepts to the table, and the farmers bring their practical experience. Together, creating a toolkit that translates permaculture into the local dialect of farming practices. Essentially, becoming partners in innovation as farmers are not just end-users, they are co-creators here.

The central aim was to develop a participatory-based permaculture toolkit together with marginal farmers to support them in moving towards sustainable farming. In other words, focusing on creating a set of tools that would make it easier for small farmers to plan and implement permaculture principles on their own land, and also to design these tools in partnership with the farmers. The objectives here included carefully documenting the design process, all the iterations, workshops, and refinements, so it can be evident to what works and what doesn't when involving farmers in tool design. The key idea was that the outcome would bridge the knowledge gap. The idea in developing the concept was to take permaculture from an abstract concept and make it something hands-on and user-friendly that a farmer with limited formal education could use in the field. By sharing this toolkit and the workshop format, this project aimed to provide a reference model for others, whether designers, NGOs, or policymakers, on how to enable marginal farmers to transition to sustainable practices.

The toolkit combines Human-Centered Design from a systemic perspective and Participatory Design as the methodological framework (Fig. 10.5) [10]. Human-Centered Design provided the systemic design process structure as it's all about empathizing with users, defining the problem, ideating, prototyping, and testing in an iterative loop while being sensitive to broader systemic insights. In the case here, the "users" were farmers, so spending a lot of time with them and understanding their needs and constraints was crucial. Then was co-creating the initial toolkit components, trying them out with farmers, gathering feedback, and refining them repeatedly.

Participatory design complemented this by ensuring the farmers themselves were, in a sense, part of the design team. As mentioned before, participatory design is an approach often used in community development, as it involves the participants

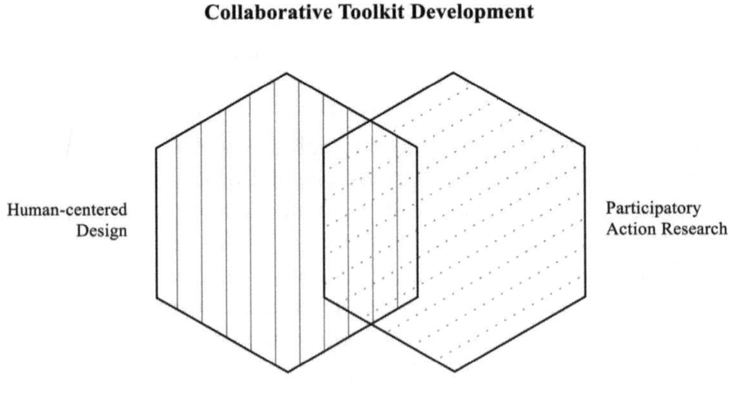

Fig. 10.5 Methodological approach of collaborative toolkit development

(here, farmers) in defining the problems, trying out solutions, and reflecting on the results. It's action-oriented and participatory, i.e., farmers had a say at each step. Through the use of visual learning tools and hands-on tools, the farmers can learn farm mapping, soil assessment, and integrated cropping systems and, therefore, graduate with sustainable agriculture easily and effectively [11, 12].

10.5 Formulating the Interventions

Further, a detailed approach was taken by working closely over an extended period with one particular farm that was in transition from conventional farming to permaculture in Ujjain, Madhya Pradesh. It demonstrated the need to understand the Design challenges in permaculture adoption in India. Here's an overview of the co-design process that has been followed, broken down into stages:

- Stage 0: Field Study & Prototype 0.
- Stage 1: Co-creation Workshops #1 & #2.
- Stage 2: Mini-Workshops #3–5.
- Stage 3: Workshop #6 (with Designers)

Stage 0: Field Study & Prototype 0, started with a foundational stage, spending a year on a transitioning farm. Through this immersion and daily farm life was observed, and initial insights were gathered. Based on the insights, the prototype (Prototype 0) of the toolkit was developed. It can be thought of as the preliminary version based on field observation and permaculture theory but not yet tested with others.

In Stage 1: Co-creation Workshops #1 and #2, Prototype 0 was taken into the real world via workshops. Workshop #1 was with a group of farmers (particularly those already practicing or familiar with permaculture) to get their ideas and opinions. Based on that, improvements were made. Next, Workshop #2 was with a group of Designers. A different perspective to examine the toolkit's design and usability was taken. Their feedback led to further refinements, resulting in Prototype 1. So, after Stage 1, the toolkit was much improved in content and usability.

Stage 2: Mini-Workshops #3–5 was conducted with Prototype 1, that included three back-to-back mini workshops with farmers in smaller groups. The idea was to observe in detail how farmers use the toolkit and gather more insights. Each mini-workshop allowed the toolkit to be iterated quickly. By the end of Stage 2, all the feedback was integrated, which further refined the toolkit.

Stage 3: Workshop #6 (with Designers) included another round of workshop with expert designers and design students. This involved a usability and design audit of Prototype 2 that focused on how easy the toolkit was to understand and use. This led to final tweaks and the addition of the digital component. A mobile application for future work, culminating in Prototype 3. By the end of this process, multiple cycles of co-creation were done, alternating between farmers and designers as collaborators to ensure the toolkit was effective and user-friendly. Going through each stage in detail of designing the toolkit, three conceptual pillars were identified that informed what

Fig. 10.6 Toolkit elements

tools the toolkit should contain and how it should function. These include Ecological Design Tools [8], Knowledge-Sharing and Learning Tools [13], and Participatory Engagement Tools [14].

The first pillar was Ecological Design Tools. These were the elements of the toolkit that help farmers apply ecological permaculture design principles on their farm. For example, tools that guide sector analysis, site analysis, and zone planning. Essentially, it included components that were needed that help to visualize and plan with farm design like simple maps or overlays showing sun paths, or markers for different farm zones (Fig. 10.6).

The second pillar was Knowledge-Sharing and Learning Tools. Permaculture involves learning new concepts that include visual aids and interactive learning materials, such as illustrated cards that explain a practice, or a workbook where farmers can draw or note down observations about their farm. These are rooted in the idea of community learning, where farmers could use them in groups to share experiences. For instance, a set of story cards might prompt farmers to discuss "What do you do to conserve water on your farm?" and share tips. The third pillar was Participatory Engagement Tools. These tools ensure the toolkit isn't a one-way teaching device but something farmers actively shape which involve co-creation exercises and feedback loops. This means including a scale model kit where farmers physically build a model of their farm, or a feedback and reflection sheet where they can note what worked or what didn't.

The toolkit was conceptualized in a way that allowed the farmers to evolve the toolkit with input and get them to articulate their tacit knowledge. Together, these three pillars, Ecological, Knowledge-Sharing, and Participatory, laid the groundwork for the types of components designed. They ensured that the toolkit was not just informative, but also interactive and co-evolving with farmer input. The idea to include visuals, tactile and farmer -friendly elements emerged by iterating the toolkit with the farmers. The Initial Prototype was the starting point, which was raw, rough, but full of intent. A base map was crafted, which was essentially a large sheet farmers could use to recreate their own land. Alongside, small illustrated cards were created which showed typical farm elements like a pond, cow shed, compost pit, mango tree, or rice crop (Fig. 10.7). The idea was that the farmers could place these elements on

Fig. 10.7 Toolkit elements as small illustrated cards

the map to explore layouts. It would work like a storytelling board, inviting farmers to narrate their land's journey and imagine what it could become. This early low fidelity version was raw and far from perfect. Some cards and elements were too abstract, some were too concrete, and farmers weren't sure what some symbols meant.

During the first workshop, one farmer had asked, "Yeh Peepal hai ya Neem?" (Is this Peepal or Neem?) Questions such as these aided in realizing the need to bring more local relevance into the visuals. At the core was a set of simple guides, no jargon, no heavy text. Just clean, easy-to-follow instructions on permaculture principles, for farmers with diverse literacy levels.

Content such as instruction manuals, design templates, interactive workbooks, visual aids, scale models, feedback, and reflection sheets were planned where initially, paper cuttings, sun boards, other elements and very rough layouts of the base were used. To further refine the understanding, during the second step, all the essential elements were developed as visuals digitally so that they can be printed and easier to replicate further. Identifying the elements and making them relevant for farmers while maintaining the scale and size of the prints was difficult. The printed cutouts were pasted on the sun board and designed in a way that they could stand on the base map without falling.

Moving into Prototype 1 (Fig. 10.8) and 2, more complexity entered, but simplicity was never compromised. In Prototype 3, the most refined version, adding something that wasn't physical at all, a digital layer. A basic mobile app interface to support the toolkit process, in cases where farmers required support [15], especially younger ones, to simulate designs, share with peers, and access stories from other farmers was designed (Fig. 10.9). Existing literature has examined zero-budget natural farming [16], participatory approaches to sustainable agriculture and participatory research [17, 18], the environmental dimensions of sustainable agriculture [19], gender and land-right issues in Indian agriculture [20], the foundations and application of participatory action research in agriculture and natural-resource management [21, 22], community resilience and Indian agricultural policy frameworks [23–25], permaculture as a strategy for developing resilient communities [26], and the use of probes, toolkits, and prototypes in co-design [27, 28].

This worked a complementary touchpoint to the physical toolkit, bridging generational and technological gaps. What began as a handful of cards and a paper map had transformed into a modular, evolving ecosystem of tools. From physical props to discussion prompts, from map overlays to digital support, each piece was shaped by the farmers themselves. The best parts of the toolkit weren't designed by the designers. They were designed through staying open, listening deeply, and letting

Fig. 10.8 Prototype 1

Fig. 10.9 Mobile app interface

the process breathe. This evolution taught that in participatory design, the toolkit isn't the end goal. It's a conversation starter, and it grows with the community it belongs to which it belongs.

Before diving into the stages, let's clarify what the Participatory Permaculture Toolkit actually is. Think of it as a "guide-in-a-box" for sustainable farming. It's not just one tool, as it's a collection of tools and materials that together help a farmer. It helps in the process of planning a permaculture farm. Here, it was deliberately made visual and interactive so that farmers could understand it easily. Traditional training might give a farmer a manual or a lecture instead; in contrast, the toolkit includes things like picture cards, simple maps and templates, and even miniature models of farm elements. This way, farmers can learn by doing, by physically arranging and visualizing aspects of their farm rather than by reading a lot of text.

The toolkit breaks permaculture down into practical activities. For example, instead of just saying "apply permaculture principles", the toolkit would have a step where the farmers map out where water flows on their land and decide where to dig a pond or a trench. Another step was designed to help them analyze zones or plans, which crops to plant together, amongst others. It covered aspects such as site observation, water management, soil improvement, crop planning, and integrating animals, which are all some key permaculture considerations, but in a very hands-on way, which is engaging and even fun. One can imagine it a like a workshop where the farmer and perhaps their family or fellow farmers gather around, use the toolkit's pieces to sketch out a farm design on a board, place little model cows or trees, discuss, and experiment with different layouts. This playful aspect lowers the intimidation factor of "planning a farm" and encourages learning by trial and error.

Using what was learnt from the experimental farm and permaculture design course, Prototype 0, the very first version of the toolkit was designed. At this stage, it was a rough prototype, that included hand-drawn sketches and makeshift materials, while encapsulating the core ideas that needed to be tested. Having identified certain tools that was useful. For instance, for the base map of the farm, a board or large sheet that could replicate a farmer's field, where things could be placed was used. Along with that, some illustration cards representing common farm elements, little pictures of different crops, trees, a well, a house, animals, etc. were also used. The idea was that the farmer could move these around on the base map to experiment with layouts as a physical simulation of their farm. So, Prototype 0 was essentially an "idea kit" as it had probes for each type of tool that could help to build a base on, visual aids, interactive pieces, and some guiding text. This nascent toolkit was taken into the real world to see how actual users (farmers and later designers) would react to it. The plan was to co-create from this baseline, show it to farmers, let them use it, and then hear what's useful, what's confusing, and what's missing.

This co-creation journey started with Workshop #1 with expert farmers. Stage 1, an interactive Co-creation Workshop, was conducted with expert farmers, specifically inviting farmers who already had some permaculture experience. About 30 farmers who had been practicing permaculture or natural farming for a few years had joined. First, the toolkit prototype to everyone was introduced to everyone, each component and how it was supposed to work was explained. The key was to get them to use the baseboard, placing the mini models and crop cards, filling in templates, etc. The facilitator observed and assisted them as needed but largely let them use the toolkit freely. The farmers were quite knowledgeable, many having their own ways and methods for permaculture. As they worked, they also shared from their own experiences. Real-time inputs on where the toolkit aligned with their thinking and where it didn't were gathered. The entire environment was similar to a collaborative brainstorming session, where the toolkit pieces were moved around, farmers sketched additional things on paper, noting down important insights.

The aim here was to gather practical feedback. Was it easy for them to use? Did it cover the important aspects of permaculture planning that they require? Did anything confuse them, or was any particular thing missing? Their insights and feedback on the toolkit's utility were important.

Workshop #1 gave valuable, practical feedback such as simplicity should be maintained as farmers liked the hands-on approach. Visuals over text was preferred for easier use by less-literate users, which hinted at icons instead of instructions. They also pointed out the need to mark the sun's direction and slope, amongst other missing features.

One farmer even tilted the board to show how slope affects water flow, and how important both features are, and the need to add them. Existing models were appreciated, but they asked for local tree icons and ways to show neighboring farms and shared resources. Also, water and soil were top priorities as they suggested adding more tools for water harvesting and soil care, like check dams and mulching. Another important insight was that some terms were too technical for the farmers to understand and recommended using local language and simpler phrasing.

After incorporating feedback from the workshops with farmers, the toolkit was further refined. This next workshop with designers brought insights from a different perspective. Eight designers and design students were split into two groups. The focus was to evaluate the toolkit's usability, visual clarity, and user experience, the things that farmers might not always point out.

Since these new participants were not from a farming background, a quick orientation was given about small-scale farming in India, including the key challenges faced by farmers, and an overview on permaculture. This helped them understand the toolkit's context and purpose. To make the activity realistic, each group received a fictional farmer persona. For example, a 45-year-old farmer with 2 acres in a drought-prone area, while another was a female farmer with separate needs and yields. The design workshop members had to use the toolkit as if helping that farmer plan their land. Each team worked through the toolkit step by step, building farm plans. A debriefing was held where they shared insights on what worked, what was confusing, and how the design could be clearer or more engaging.

This workshop helped to fine-tune the visual design, layout, flow, and user engagement, bringing the toolkit one step closer to being truly user-friendly. Designing with people, not just for them, resulted in richer, more relevant results. Entering this process, typically, the designers believed that they were here to "teach" or "assist." But in reality, the designers were the learners too, as the workshops showed how critical it is to design for context, especially when working with users who may not be literate or have prior exposure to design thinking.

10.6 Enabling the Transition

The toolkit had to speak through visuals, through touch, and through storytelling. Through every iteration, farmers became more invested, not only in the toolbox but also in the process. Their ideas shaped the toolkit, and that ownership led to deeper engagement, realizing that permaculture itself must be localized. Although its ethics and values are universal, their application varies by area, soil, community, and culture. So after many rounds of fieldwork, feedback, and redesign, the final toolkit was developed, which was the most refined version of the toolkit (Fig. 10.10).

The toolkit is not just a guidebook or a set of instructions. It's akin to a hands-on, interactive planning kit that farmers can actually use to visualize and redesign their farm. There's a large modular board that represents their land, and they can place moveable cards like crops, trees, water paths, animals exactly where they think they should go. It also includes simple worksheets for things like mapping the sun's path, testing the soil, or planning zones.

To make the experience more relatable, story-based learning cards were added that share real-life scenarios from other farmers. Kind of personas, inviting users to reflect and apply lessons to their own context. One of the newest additions is a digital version, a mobile-friendly prototype that would help farmers, especially the younger generation. In order to design the farmland, the idea behind this toolkit was always the same: keep it visual, flexible, and something that sparks conversation and

Fig. 10.10 Refined toolkit

learning. The focus is not just the individuals, but within the farming community. In early tests, something interesting was observed as the farmers stayed engaged for hours, sharing ideas and testing layouts.

A significant observation was that how the toolkit became a conversation catalyst. Farmers weren't just designing farmland, they were debating, explaining, and storytelling. A few even mentioned they had never thought about farm zones or sun patterns until they visualized them through the toolkit. Peer-to-peer learning emerged, where farmers showed each other how they did things, turning the toolkit into a community learning asset. This kind of response for the toolkit has the potential to evolve beyond just a design tool; it could become a medium for ongoing knowledge exchange. What does this mean beyond this project? How can one truly think about participatory design beyond this?

10.7 Conclusion

Looking at the toolkit and its advantages, first, it offers a model for agricultural extension rooted in co-creation. If governments or NGOs adopt participatory tools such as these, the impact could be transformative, especially in areas where farmers are underserved by existing training systems. Second, this toolkit supports climate resilience. Permaculture's regenerative principles can help against drought, pest outbreaks, and chemical dependencies. Third, there's scalability. With cultural and regional modifications, this model can be adapted to other states or countries with smallholder challenges. Finally, the fourth one, as designers, this journey has deepened the understanding of rural design pedagogy, how to teach, learn, and create with rural communities in ways that are empowering, dignified, and inclusive.

This project would not have been possible without the farmers who opened their homes and minds to this journey. Their trust and participation were the foundations, as it reminded them of the deeper purpose of systemic design for social innovation, which is not just to create beautiful objects, but to create agency. When designed with empathy, curiosity, and respect, it's not just about building tools, but building communities of possibility. Let's continue to ask, who are the designers designing for? Better yet, who are they designing with?

References

1. Faludi J, Acaroglu L, Gardien P, Rapela A, Sumter D, Cooper C (2023) Sustainability in the future of design education. She Ji: J Des Econ Innov 9(2):157–178
2. Pingali P (2012) Green revolution: impacts, limits, and the path ahead. Proc Natl Acad Sci USA 109(31):12302–12308
3. Reddy D (2003) Organic farming and sustainability in India. Centre for Sustainable Agriculture
4. Chambers R (1994) The origins and practice of participatory rural appraisal. World Dev 22(7):953–969

5. Mollison B (1988) Permaculture: a designer's manual. Tagari Publications, Tyalgum
6. Sharma A, Sharma N (2023) Design challenges in permaculture adoption in India: case study of wheat and soya bean. In: Chakrabarti A, Singh V (eds) Design in the era of industry 4.0, vol 2. ICORD 2023. Springer, Singapore
7. Ferguson RS, Lovell ST (2014) Permaculture for agroecology: design, practice, movement, and worldview. Agron Sustain Dev 34(2):251–274
8. Holmgren D (2002) Permaculture: principles and pathways beyond sustainability. Holmgren Design Services, Hepburn
9. FAO (2017) The future of food and agriculture: trends and challenges. FAO, Rome
10. Baum F, MacDougall C, Smith D (2006) Participatory action research. J Epidemiol Community Health 60(10):854–857
11. Pretty J (1995) Participatory learning for sustainable agriculture. World Dev 23(8):1247–1263
12. Pretty J (2008) Agricultural sustainability: concepts, principles and evidence. Philos Trans R Soc Lond Ser B Biol Sci 363(1491):447–465
13. Wenger E (1998) Communities of practice: learning, meaning, and identity. Cambridge University Press, UK
14. Whyte WF (1989) Advancing scientific knowledge through participatory action research. Sociol Forum 4(3):367–385
15. Sharma A, Malhotra S, Kant V (2025) Resilience through participatory design: promoting sustainable agriculture among Indian marginal farmers. In: Chakrabarti A, Singh V, Onkar PS, Shahid M (eds) Responsible and resilient design for society, vol 1. ICoRD 2025. Springer, Singapore
16. Budget Z (2022) Zero budget natural farming. Res Today 4(3):186–189
17. Bruges M, Smith W (2008) Participatory approaches for sustainable agriculture: a contradiction in terms? Agric Hum Values 25(1):13–23
18. Cornwall A, Jewkes R (1995) What is participatory research? Soc Sci Med 41(12):1667–1676
19. Etingoff K (2019) Sustainable agriculture and the environment. Apple Academic Press, Palm Bay
20. FAO (2018) Gender and land rights in Indian agriculture. https://www.fao.org/gender/learning-center/thematic-areas/gender-and-land-and-water/en
21. Fals-Borda O (2001) Participatory action research in social theory: origins and challenges. In: Reason P, Bradbury H (eds) Handbook of action research: participative inquiry and practice. SAGE Publications, Thousand Oaks, pp 27–37
22. Gonsalves J, Becker T, Braun A, Campilan D, De Chavez H, Fajbe E, Kapiriri M, Rivaca-Caminade J, Vernooy R (eds) (2005) Participatory research and development for sustainable agriculture and natural resource management: a sourcebook, vol 1–3. International Potato Center, Laguna
23. King CA (2008) Community resilience and contemporary agri-ecological systems: reconnecting people and food, and people with people. Syst Res Behav Sci 25(1):111–124
24. Ministry of Agriculture & Farmers Welfare (2020) Agriculture census report. Government of India
25. NMSA (2015) Guidelines for promoting sustainable agriculture in India. https://nmsa.dac.gov.in
26. Olmos Carbonell E (2023) Permaculture: contributing to sustainable development by designing resilient communities
27. Sanders EB-N, Stappers PJ (2014) Probes, toolkits and prototypes: three approaches to making in codesigning. CoDesign 10(1):5–14
28. Sanders EB-N (2012) Convivial toolbox: generative research for the front end of design. BIS Publishers, Amsterdam

MIX
Papier aus verantwortungsvollen Quellen
Paper from responsible sources
FSC® C105338

If you have any concerns about our products,
you can contact us on
ProductSafety@springernature.com

In case Publisher is established outside the EU,
the EU authorized representative is:
Springer Nature Customer Service Center GmbH
Europaplatz 3, 69115 Heidelberg, Germany

Printed by Libri Plureos GmbH
in Hamburg, Germany